Thunderstorms, Tornadoes, and Building Damage

Thunderstorms, Tornadoes, and Building Damage

Joe R. Eagleman
Vincent U. Muirhead
Nicholas Willems
University of Kansas

Lexington Books
D.C. Heath and Company
Lexington, Massachusetts
Toronto   London

Library of Congress Cataloging in Publication Data

Eagleman, Joe R
   Thunderstorms, tornadoes, and building damage.

   Bibliography: p.303.
   Includes index.
   1. Tornadoes.　2. Thunderstorms—United　States.　3. Building
failures. I. Muirhead, Vincent U., joint author. II. Williams,
Nicholas, joint author. III. Title.
QC955.E33　　　690　　　74-30674
ISBN 0-669-98137-0

Published simultaneously in Canada

Printed in the United States of America

International Standard Book Number: 0-669-98137-0

Library of Congress Catalog Card Number: 74-30674

Contents

List of Figures

List of Tables

Preface

Tornadoes are a source of injury to several thousand people in the United States every year. As the population increases, the probability of tornadoes striking cities increases, because of their larger size. Large sections of many cities are inadequately constructed and have no provisions for shelter from severe storms. Little thought has been given in the past to the design of tornado resistant dwellings or to safety factors in existing buildings.

This book has been developed as a result of investigations following a major tornado in Topeka, Kansas, in 1966 and subsequent research conducted at the University of Kansas. Part of these efforts were supported by Environmental Control Administration, Department of Health, Education, and Welfare. Several different aspects of tornadoes have been investigated each directed toward improving our understanding of tornadoes and our ability to decrease their impact on society.

Careful observations have been made of several hundred houses damaged by tornadoes. In addition, it has been apparent from damage observations that definite wind flow patterns must have existed during the tornadoes. These patterns have been modeled in the laboratory by developing the first traveling tornado-like vortex by a combination of updraft and circulation of air. Application of the theory of airflow around a solid cylinder has led to the development of a new theoretical thunderstorm model. This model gives additional insight into such illusive concepts as the source of energy for the tornado and the internal characteristics of severe thunderstorms. The tornado vortex has been analyzed as a compressible flow phenomenon. An index has been developed for forecasting tornadoes based on energy considerations and the wind-shear profile required for the double vortex thunderstorm model. Various types of structural tests on building components have been conducted in the laboratory. Pressure mapping for various building designs has been completed in the wind tunnel. Destruction model testing of buildings has given additional information about structural weaknesses and venting effects on the ability of buildings to withstand high wind velocities. All these various facets are combined in the last two chapters.

It is hoped that the various studies of thunderstorms, tornadoes, and building damage have significantly advanced our understanding of the severe weather component of our environment, and, perhaps, have placed us in a better position for thinking in terms of controlling this portion of our environment.

The authors gratefully acknowledge the assistance of graduate and undergraduate students during the past several years while these investigations have been in progress. The following students have been associated with the faculty investigators as research assistants; Robert C. Brown, Larry D. League, Phillip L. Garner, Robert Shields, Audie Chen, Marvin Stone, Issac Allotey, James F. Iams, Wen C. Lin, Donald Perkey, James Shortridge, Tianlai Hu, Hampton N. Shirer, and Charles Royer. The encouragement of our wives has been gratefully appreciated in addition to the assistance of Doris L. Eagleman in typing the manuscript.

Thunderstorms, Tornadoes, and Building Damage

1

Characteristics and Detection of Tornadoes

Tornado Characteristics

Tornadoes are the most violent of all the vortex storms that occur in the atmosphere. They are much stronger than their smaller cousins, dust-devils, and their two larger cousins, hurricanes and mid-latitude cyclonic storms. Although the strongest dustdevils and hurricanes may surpass the intensity of the weakest tornadoes, the intensity of the average tornado is unmatched.

The strongest recent hurricane, Camille in 1969, contained wind velocities slightly greater than 200 mph. The speed of tornado winds and the low pressure within the core have not been measured accurately. Tornado wind speeds estimated or calculated from various structural damage give speeds from 65 to 500 mph [1, 2].

The loud noise accompanying tornadoes, frequently described as the roar of many jets or trains, must have a source within the tornado or lower part of the thunderstorm to which it is attached, since the noise is present with funnel clouds that do not touch the ground. The source of the loud noise may well be supersonic winds within the tornado.

The typical midwestern tornado lasts for about 30 minutes, covers a distance of about 15 miles, and leaves a damaged strip about 300 yards wide. Occasionally a tornado will stay on the ground for over 200 miles and the largest ones leave a damage path greater than one mile wide.

A record number of tornadoes occurred in 1973 when 1,109 tornadoes were reported in the United States causing 87 deaths. The average number of tornadoes is 708 per year with 93 deaths and many times this number of injuries.

Tornadoes are generated by the largest thunderstorms (cumulonimbus clouds). A typical tornado day starts with very humid and warm air at ground level and may be free of clouds during the morning hours. As the day progresses solar heating of the ground combined with several identifiable atmospheric conditions set the stage for the development of buoyant air, which results in isolated large cumulonimbus clouds (Figure 1-1) or lines of cumulonimbus (squall lines). The severe thunderstorm may extend to 55,000 feet above the surface with a different internal structure from the average, ordinary thunderstorm. Gusty winds, large hail, and tornadoes are produced by these large severe thunderstorms.

1

Figure 1-1. Cumulonimbus Clouds in Various Stages of Development from Mature (Left) to the Initial Developing Stage. View is toward the south from Lawrence, Kansas. (Photograph by J. R. Eagleman)

Appearance of Tornadoes

Development and Visibility

Tornadoes are rotating vortices that extend to the ground from the base of a thunderstorm. If the cloud base is too high or if the vortex is too weak to reach the ground it is designated a funnel cloud. A vortex is composed of intense, circulating air around a central core. Most tornadoes circulate cyclonically (counterclockwise when viewed from above) in the Northern Hemisphere. The exact nature of the winds in a tornado have not been measured and are, therefore, subject to speculation. The cyclonic winds spiraling around the core vary in intensity and attempts have been made to assign a scale to tornadoes on the basis of types of damage [3]. Even less information is available on the central core of the tornado. It has been suggested by some [4,5] that the tornado core is composed of a downdraft similar to the dustdevil and hurricane. However, tornado damage investigations and laboratory modeling experiments indicate that it is more likely composed of an updraft core of considerable velocity.

3

Figure 1-2. Mammatus Cloud Formation That Frequently Accompanies Severe Thunderstorms. Bulging pouches indicate vertical motion. A funnel cloud developed from this thunderstorm over Russell, Kansas. (Photograph courtesy of Ellen Grass)

A tornado is spawned from a large thunderstorm that has developed the proper internal structure of updrafts and circulation. Very large thunderstorms with intense lightning, hail, and mammata formation (Figure 1-2) may be capable of producing tornadoes. A funnel cloud forms in the southwestern part of the base of the cumulonimbus cloud and begins to drop toward the ground. As it develops and descends from the cloud base it is normally funnel shaped. This sequence was photographed on May 11, 1970, near Ellsworth, Kansas (Figure 1-3). The funnel cloud is visible because of condensation in the lower pressure within the vortex. Water vapor within the funnel condenses into cloud droplets as the air expands and cools because of the lower pressure. This is the same process that causes cloud formation in general, except that the air normally has to be lifted to higher elevations where the pressure is lower and condensation can occur because of expansion and cooling.

As the funnel cloud touches the ground and becomes a tornado its appearance is affected by dust, water, or debris that is picked up from the surface. The tornado in Figure 1-3 is visible because of both dust and cloud droplets in the last stage photographed, as well as in some of the previous stages. The central funnel shape is composed of droplets while the outer layers contain material picked up from the surface.

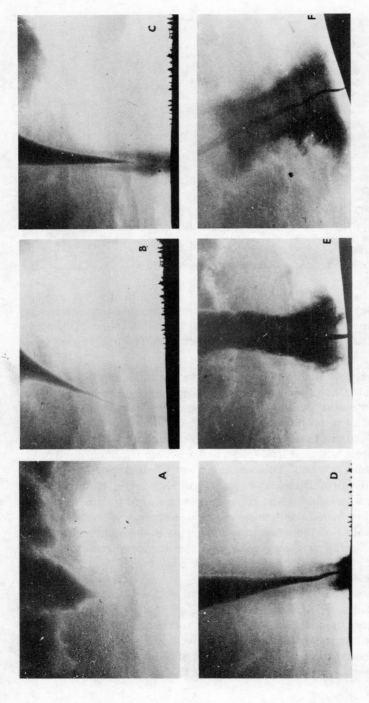

Figure 1-3. Development of a Tornado Near Ellsworth, Kansas, on May 11, 1970. The vortex may be present but invisible below the funnel extending from the cloud until it picks up surface material. (Photographs by Kansas Highway Patrolman Vernon Davis, courtesy of Topeka National Weather Service)

Size, Shape, and Color

At the ground the typical tornado is about 300 yards in diameter, but their size has ranged from only a few yards to more than a mile wide. The vertical length of tornadoes is governed by the height of the parent cloud base. The cloud base height is related to the degree of saturation of the air or the difference between air temperature and dew point temperature. Cloud base may range from a few hundred feet to several thousand feet. Thunderstorms with low bases are more likely to have stronger tornadoes if they also have sufficient vertical development and the proper internal structure.

Tornadoes develop in a variety of shapes as well as sizes. The edges of the tornado may be very irregular or quite smooth. Figure 1-4 shows a funnel with irregular edges. If the funnel is smooth it can take a variety of shapes, from the typical funnel that narrows toward the ground (Figure 1-5) to the cylindrical shape that is almost constant with height (Figure 1-6). A single tornado may first appear with an irregular or smooth funnel shape and later transform into a more cylindrical, smooth shape, followed by a rope shape (Figure 1-7) before dissipating. Sometimes the central invisible core of the tornado is outlined by the cyclonic winds rotating around it (Figure 1-8). This happens when the amount of water vapor and dust within the tornado is small. Occasionally a tornado has unusual bends as in Figure 1-9.

The color of a tornado depends on the background against which it is viewed and the material incorporated from the surface. The tornado funnel is normally located in the southwest part of the thunderstorm with the major rain area in the northeast part of the thunderstorm. If the tornado is viewed looking toward the northeast with the dark rain behind the funnel it may appear white (Figure 1-10). The same funnel viewed toward the southwest with clear skies behind it may appear very dark (Figure 1-11). In addition to variations in grey tones because of the background and amount of condensed water in the funnel, dust and debris from the surface cause a wide variation in colors. Depending on the type of soil beneath it, dust can turn a tornado into odd shades of brown, yellow, and red.

Electrical charges within the funnel can also contribute to eerie effects. Funnels have been reported to glow on the outside as well as from the inside. An eye witness description of the appearance of the inside of the funnel [6] indicates that it is charged with electricity and is very smooth because of the high velocity winds.

Multiple Tornadoes

More than one tornado can be supported by the same thunderstorm up-

Figure 1-4. Tornado with Irregular Edges Characteristic of Large, Intense Tornadoes. This tornado traveled across Topeka, Kansas, on June 8, 1966, causing 16 deaths, 406 injuries, and over $100 million damages. (Photograph courtesy of Topeka Capital-Journal)

Figure 1-5. Funnel Shaped Tornado (Photograph courtesy of National Oceanographic and Atmospheric Administration)

draft. These can be small, smooth funnels or quite large ones. Two large funnels were photographed during the Palm Sunday tornado outbreak, April 11, 1965. One of these rotated around the other [7], showing the larger circulation of the parent cloud.

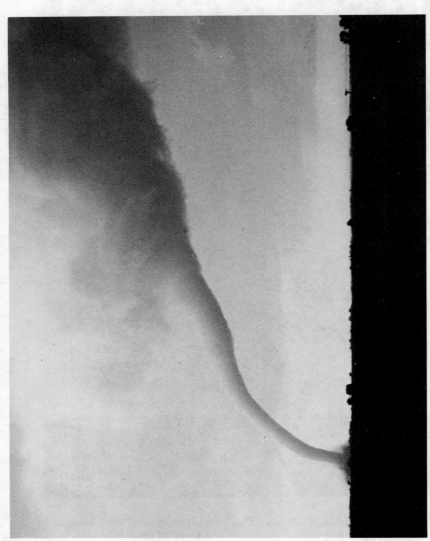

Figure 1-6. Cylindrical Curved Tornado That Also Shows the Tendency for Tornadoes to Approach the Ground from a Perpendicular Position as Much as Possible. This tornado struck Tracy, Minnesota, on June 13, 1968, resulting in 9 deaths and 125 injuries. (Photograph by Eric J. Lantz of the Walnut Grove, Minnesota Tribune)

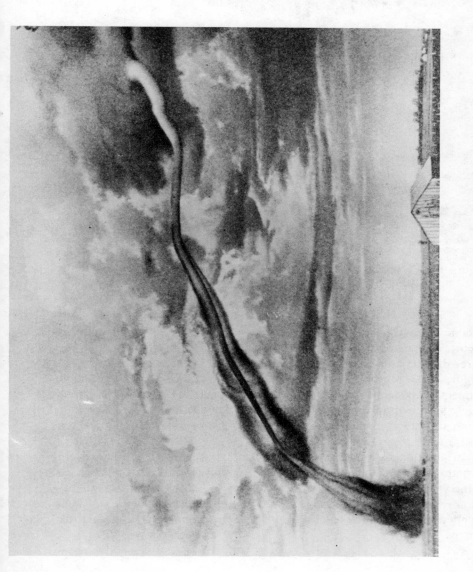

Figure 1-7. This Classical Photograph Shows the Rope Shape That is Frequently the Last Stage of the Tornadoes Life Cycle (Photograph courtesy of National Oceanographic and Atmospheric Administration)

Figure 1-8. Tornado with Visible Hollow Core Surrounded by Rotating Winds That Occurred Near Kingfisher, Oklahoma, in August 1965 (Photograph by Wayne C. Carlson)

Photographs of multiple tornadoes are not common, although there are reports of as many as five or six funnels extending from one thunderstorm. Three distinct funnels were photographed in 1973 near Marquette, Kansas (Figure 1-12). Multiple funnels probably rotate in the same direction and connect to the same updraft within the thunderstorm. A series of photographs is shown in Figure 1-13 of multiple funnels that occurred in Randolph County, Indiana, on April 3, 1974. These illustrate the unstable nature of multiple funnels.

In addition to multiple funnels from a single thunderstorm, several tornado producing thunderstorms may be within a radius of 50 miles or less at the same time. This occurs primarily on days with major tornado outbreaks such as April 11, 1965, or April 3, 1974. More frequently, a single thunderstorm produces several tornadoes, one originating as another dissipates.

Distribution of Tornadoes

Tornadoes have occurred in every state in the United States including Alaska and Hawaii. However, they are much more frequent in the Great

Figure 1-9. Tornado with Two Relatively Sharp Bends in the Vortex as Photographed on April 11, 1965 (Photograph courtesy of Nicholas Polite, Wanatah, Indiana)

Plains region east of the Rocky Mountains. The central United States is susceptible to more frequent tornado activity because of its unique geographical setting. Several specific atmospheric characteristics are required for tornado development. An unstable atmosphere is one of the requirements for large thunderstorm development. The atmosphere becomes less stable and more susceptible to the development of localized bubbles of buoyant air as a result of two different processes. One of these arises because of solar heating of the ground and lower atmosphere, with the result that tornado activity reaches a peak from 4 to 6 P.M. (Figure 1-14) because of this heat input and increased buoyancy of the lower atmosphere. This process is not unique to the central United States, however.

The second process that leads to decreased atmospheric stability occurs on a larger scale. Tornado outbreaks are associated with the flow of cool, dry air over the Rocky Mountains, where it descends into the Great Plains region. At the same time warm, moist air is flowing northward from the Gulf of Mexico. During days with tornado activity there is frequently a sharp boundary in the troposphere where the cool, dry air from the west is overriding the warm, moist air from the south. This has the same effect on atmospheric stability as solar heating since the stability of the atmosphere is reduced and it is easier for thunderstorms to develop. Therefore, because of its geographic setting with the mountain barrier to the west and the Gulf

Figure 1-10. Sometimes the Tornado May Appear White as in This Photograph of the Union City, Oklahoma, Tornado of May 24, 1973. This happens when the darker rain area is in the background. (Photograph by Robert Gannon, courtesy of National Severe Storms Laboratory)

to the south or southeast, the central United States has the greatest tornado activity (Figure 1-15).

Tornadoes are known to be associated with the jet stream. The jet stream is a high velocity stream of air, flowing from west to east at different latitudes during different seasons of the year. Although it frequently meanders like a river and is not always at exactly the same location for any given month of the year, its position shifts northward during the summer months and southward during the winter (Figure 1-16). The jet stream normally

13

Figure 1-11. Another Photograph of the Union City, Oklahoma, Tornado with Lighter Skies and No Rain in the Background Shows the Same Funnel That Now Looks Very Dark (Photograph by Robert Gannon, courtesy of National Severe Storms Laboratory)

Figure 1-12. Three Funnels from the same Thunderstorm Occurred Near Marquette, Kansas, on September 23, 1973 (Photograph courtesy of Stanley Engdahl)

separates the cold air to the north from the warm air to the south. With the jet stream flowing through the southern United States in January, tornadoes are more likely in the southern states. As the jet stream moves northward during the spring months the region of tornado activity also moves northward. For the United States as a whole, more tornadoes occur in May than any other time of the year (Figure 1-17). Although the number is quite low during the winter months, tornadoes occur during all months of the year with a distinct peak in number of tornadoes in late spring.

Tornadoes are known to occur in other parts of the world more frequently than is commonly assumed. Australia is second to the United States in terms of tornado frequencies [3]. Even though it is almost as large as the United States the frequency is only about 15 per year in comparison to about 700 per year for the United States. Tornadoes throughout the world are concentrated in belts from about 20° to 50° on both sides of the equator (Figure 1-18). These belts correspond to the zones of the jet stream and the presence of contrasting air masses.

The average number of deaths from tornadoes in the United States is 93 per year based on statistics from 1916 to 1973. This number has decreased significantly in recent years due to better warning procedures. The average

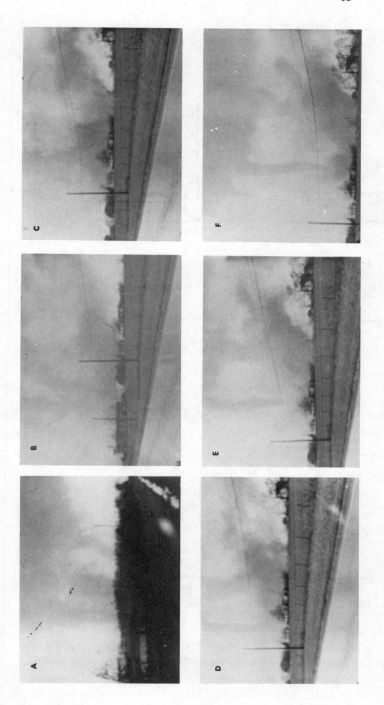

Figure 1-13. Multiple Funnels That Crossed Randolph County, Indiana, on April 3, 1974 (Photographs by Wally Hubbard, WISH-TV News, Indianapolis, Indiana)

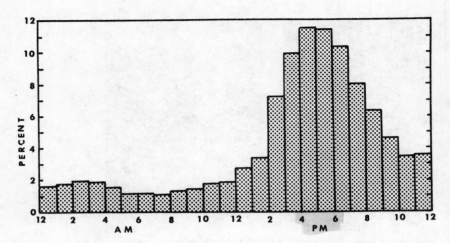

Figure 1-14. Hourly Distribution of Tornadoes in the United States Based on 4,679 Tornadoes from 1916-50 (Weather Bureau, Technical Paper No. 20 [25])

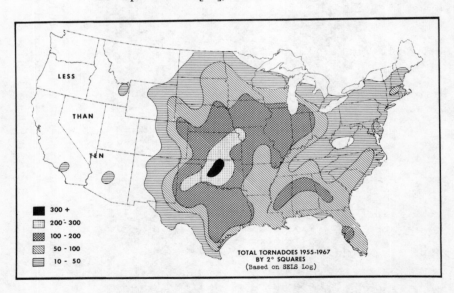

Figure 1-15. Distribution of Tornadoes in the United States (M. E. Pautz [58])

number of deaths per year was 227 from 1916 to 1950, with a peak in 1925 when 794 people died from tornadoes. Of this number, 689 deaths resulted from a single tornado that traveled over 200 miles through Missouri, Illinois, and Indiana. More than 2,000 people were injured by this tornado in

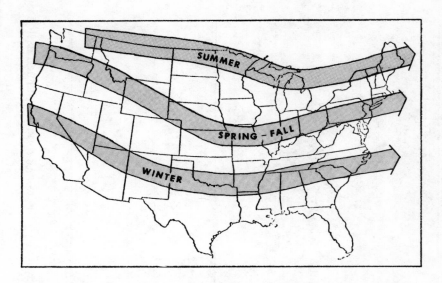

Figure 1-16. The General Position of the Polar Jet Stream During Different Seasons of the Year. Location of tornado activity is related to the jet stream.

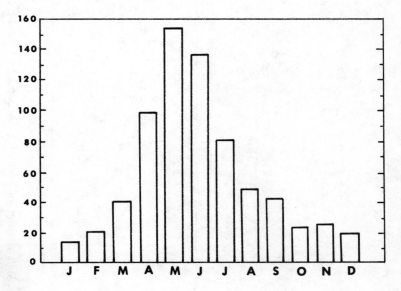

Figure 1-17. Monthly Distribution of Tornadoes in the United States Based on 12,739 Tornadoes from 1956-73 (National Oceanographic and Atmospheric Administration, Climatological Data, National Summary, Annual 1973 [23])

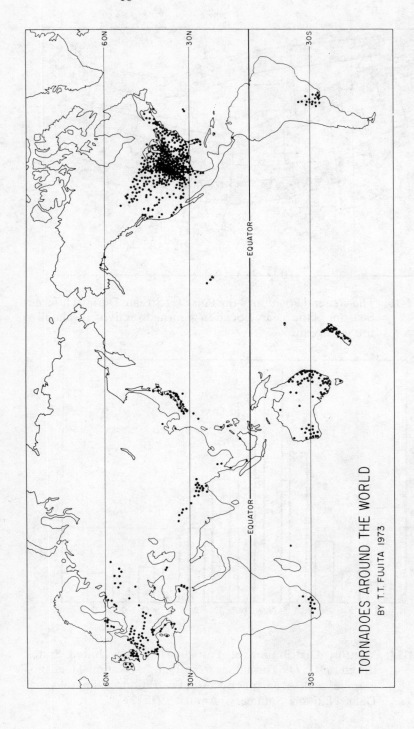

TORNADOES AROUND THE WORLD

BY T.T. FUJITA 1973

Figure 1-18. Distribution of Tornadoes Around the World Expected per Four-year Period (Reprinted with permission of T.T. Fujita)

1925. The current injury rate is not insignificant, amounting to several thousand per year.

Two tornadoes have caused over 100 million dollars worth of damage in single cities. These were the tornadoes in Topeka, Kansas, in 1966 and in Lubbock, Texas, in 1970. The most devastating single-day onslaught of tornadoes occurred on April 3, 1974, when tornadoes struck in 11 midwestern and southern states. Total damages were about $700 million, with 329 deaths and more than 4,000 injuries.

Tornado Detection

Forecasting and Detection Methods

Before the first attempted tornado forecast in 1948, a family unfortunate enough to be in the path of a tornado was forced to depend on several unreliable and inaccurate methods for advance warning of an approaching tornado. There are numerous reports of warnings provided by the abnormal behavior of farm animals and pets prior to the arrival of a tornado. Dogs seem to be particularly sensitive to tornadoes that are some distance away. Perhaps their ears are sensitive to noises produced by tornadoes that are inaudible to human ears. Other primitive methods of tornado detection, which provided several precious minutes of advance warning, included careful observation of the color of clouds with a change to an ominous greenish hue indicating severe weather. Another warning signal was the tremendous noise of the tornado. This noise has been described as a deafening roar similar to hundreds of trains or jets, or as the hum from a swarm of bees large enough to darken the sky. The associated noise is still sometimes the first warning of an approaching tornado even in this day of modern technology.

As population increased, eyewitness reports increased proportionately, and people now have greater opportunity for forewarning through modern communication networks and the efforts of groups of local "storm spotters" and highway patrolmen. Technological advancement made it possible to issue the first public tornado forecast in 1948, which eventually led to the creation of the National Severe Storms Forecasting Center (NSSFC). This center located in Kansas City, Missouri, is responsible for issuing tornado forecasts for the entire United States. Today if NSSFC issues a tornado forecast, 40 percent of the time a tornado will actually occur. Tornado forecasts are based on the proper atmospheric conditions for tornado development and, therefore, apply to areas of several thousand square miles. Tornadoes occur when a combination of atmospheric variables coincide, such as an unstable atmosphere, upper air divergence, upper air inversion layer, a high dew point temperature, a proper wind profile, and the jet stream. There is normally also a squall line, and a mid-latitude cyclone approaching. Tornado warnings for much smaller areas than the large tornado watch areas are possible only after visual sighting or detection by radar, TV, or experimental electronic detectors.

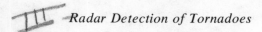

Radar Detection of Tornadoes

Since tornadoes develop from thunderstorms, the detection of thunderstorms is an important first step. This can be done by the use of radar. The electromagnetic waves of radar bounce off the moisture drops of the cloud. This gives information on the size of the storm, since the moisture drops farther away in the cloud return the wave later. The intensity of the signal also gives an indication of the severity of the storm. From radar, therefore, the size, position, direction, and speed of the storm may be determined. Tornadoes are too small and have an insufficient amount of water in the funnel to show up on the radar screen. However, thunderstorms containing tornadoes have a unique hook shape (Figure 1-19). The hook echo can be seen only if the radar antenna is properly positioned, since it is characteristic of the lower part of the thunderstorm cell and cannot be seen if the antenna is pointed to the top or middle part of the cell. It is unusual to have several thunderstorms with tornadoes in action within range of a single radar as shown in Figure 1-19. This activity occurred in the Cincinnati area on April 3, 1974. Figure 1-19 shows two thunderstorm cells that have the hook shape near their base indicating the proper structure inside the thunderstorm for tornado development and support.

The first hook echo indicating tornado activity was photographed in Illinois in 1953. Since that time several advances have been made. It is now possible to use Doppler radar to gain some information on airflow patterns inside thunderstorms by using precipitation particles as tracers of the wind [8,9,10]. Although small-scale airflow within a tornado funnel cannot be identified, the hook-shaped echo and echo-free vault corresponding to the internal structure of the thunderstorm cell can be seen, in addition to information gained on circulation within the storm. Even so, we still do not have complete information on the internal structure of tornado producing thunderstorms, since some very important regions are either too small or do not have enough precipitation in them to be visible by Doppler radar.

The National Severe Storms Laboratory in Norman, Oklahoma, and others have used digital radar for investigating thunderstorms. Weak echo regions were observed in all thunderstorms that had tornadoes [11]. It was concluded that these regions corresponded to updrafts within the thunderstorm.

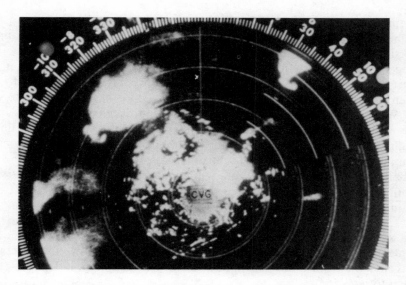

Figure 1-19. Hook Echoes Obtained by the Weather Radar Near Cincinnati on April 3, 1974. These echoes represent a horizontal slice through the lower part of the clouds and reveal shapes indicative of tornado development and support.

The hook-shaped echo of tornado producing thunderstorms is of considerable value in tornado detection. Although there is no certainty that a tornado is actually on the ground when a hook-shaped echo appears, a sufficiently strong relationship exists that many TV stations operate weather radar equipment for severe storm detection.

Tornado Detection by Television Sets

Unusually intense lightning and other electrical phenomena are often associated with thunderstorms producing tornadoes. Electromagnetic energy is generated at a variety of frequencies. Storm energy in the low frequencies of the spectrum is due primarily to electrical strokes and, therefore, could not be used to specify tornadic activity as may be possible with other frequencies.

A promising method of detecting tornadoes began in 1968 as a result of several years of developing and testing by Newton Weller [12]. It is based on tuning a television set to channel 13, darkening, then tuning to channel 2 where a brightening of the screen may indicate a tornado is near. When George Vogel heard of this new method on the first day that it was released, while attending a going-away party in Orange City, Iowa, on September 22, 1968, he was skeptical [13]. Even though Orange City's tornado alert (a beeping fire truck) sounded its admonition and the television of his host's home verified the proximity of a tornado, the anything-but-severe weather outside seemed to justify passing off these warning signs as shams. George Vogel did not believe that a tornado could be detected with a television, at least not until a telephone call informed him that his new house had just been swept away by a tornado.

Weller first began tracking storms with an oscilloscope and its one horizontal line until he realized that a television with a raster of 525 lines might be a better instrument. Weller's method will operate with any television, color or black and white, tube or transistor, with outside antenna or rabbit ears. Weller maintains that with an average television an observer familiar with its performance is capable of detecting a tornado within 15 to 20 miles, thus allowing time to take shelter.

Channel 2 is a 55 megacycle band and is the television channel closest to the electromagnetic frequency of the pulse of a tornado. It is thought by some that all tornadoes emit electromagnetic energy in different wavelengths from a normal thunderstorm with lightning, but this is still open to question.

The Weller method consists of the following steps:

1. Turn on your television set and let it warm up.
2. Turn to channel 13. Using the brightness control knob, darken the screen so it is almost black.
3. Turn to Channel 2. Leave the set alone. Do not reset the brightness after the initial adjustment.
4. Lightning appears on the screen as horizontal streaks or flashes. (A color TV produces colored lightning.) As long as the screen does not have a steady glow, the storm is not a tornado.
5. The signal of a tornado is an increasingly steady, bright, white light. Or, if there is a station in your area on channel 2 and the darkened picture becomes visible and remains visible, a tornado is coming.
6. Take shelter. Do not get so carried away in watching the screen that you forget to seek cover—fast.

Experimental Electromagnetic Tornado Detectors

The tornado funnel is highly electrified, and it is thought that there is continuous electrical discharge within most funnels, which accounts for the glow associated with some tornadoes reported by eyewitness observers. Research at Iowa State University [14] indicates that tornadoes emit energy at 1 to 53 megahertz. Several different types of pulsed energy exists during tornadoes. It is thought that the majority, but not all, tornadoes emit certain types of electromagnetic pulsations that are different from those emitted by ordinary thunderstorms.

Research is now in progress at the Wave Propogation Laboratory in Boulder [15], which is designed to test the effectiveness of receivers operating at 3.16 megahertz for tornado detection purposes. Twenty portable instruments were placed in operation at different locations during 1974. These were equipped with directional antennas, while tests in prior years were conducted with omnidirectional antennas. During 1972 tornado detectors were installed at 18 locations in the midwestern United States. Two amplitude threshold levels were used corresponding to nominal ranges of 30 km to 70 km for typical sferics. The electrical activity of thunderstorms was characterized in terms of burst rates. To be counted as a burst of energy, the signal was required to exceed an amplitude threshold level and also to exceed a rate of 500 per second for more than 0.1 second. Burst rates exceeding 3, 10, 20, and 30 per minute were recorded. Burst rates corresponding to various weather events are shown in Figure 1-20. The tallest blocks in each weather event category correspond to average burst rates for the long range sensitive channel measuring activity out to 70 km, with the shortest blocks corresponding to the electromagnetic activity generated less than 30 km away. Using a burst rate of 20 per minute as an indication of a tornado within range of the detector, 73 percent of the reported tornadoes would have been identified by this method. Local nonsevere thunderstorms can also produce a high burst rate, however. The 1972 data indicated that over one-half of the tornado alarms arose from local nonsevere thunderstorms. The false alarm rate was improved during the 1973 season giving some indication that electromagnetic detection techniques may be perfected in the future that will be of great value in the advance warning of tornadoes.

Control of Tornadoes

The tornado is so destructive to objects in its path that it is only natural to hope that it can be controlled or modified by modern science and technology. Current understanding of the thunderstorm and tornado generation process is not sufficiently developed to this point, however.

A current research effort by NOAA's Environmental Research Laboratories in Boulder, Colorado, that has promise is based on the importance of solar heating of the ground during hot humid days when tornadoes are likely. It has been observed that jet aircraft leave contrails that spread and form high cirrus clouds under the right conditions. Perhaps this can be used to advantage by creating a cirrus cloud deck on days when tornadoes could develop. This would decrease surface heating and make it less likely

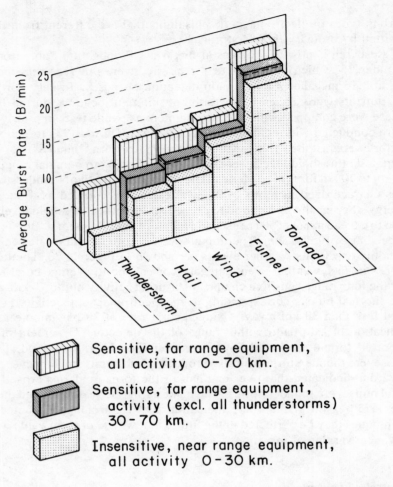

Sensitive, far range equipment,
all activity 0 - 70 km.

Sensitive, far range equipment,
activity (excl. all thunderstorms)
30 - 70 km.

Insensitive, near range equipment,
all activity 0 - 30 km.

Figure 1-20. The Number of Burst Rates for Different Weather Phenomena Measured by an Electro-magnetic Tornado Detector (Reproduced with permission from W. L. Taylor [15])

for volumns of air to become buoyant enough for severe thunderstorm formation.

Whether this or other ideas are successful, tornado control is not a current operational capability. If we cannot control or modify tornadoes, we can decrease their impact on society by accurate forecasts and warnings and by obtaining as much information as possible on the nature of tornado damage in urban areas. If certain locations are safest in houses and if houses can be designed to withstand higher winds this information should be utilized to advantage. The following chapters describe statistical and structural studies of tornado damaged houses along with related tests of model houses and theoretical considerations of thunderstorms and tornadoes aimed at advancing our knowledge of these important atmospheric phenomena.

2

Tornado Damage Patterns in Topeka, Kansas

On June 8, 1966 [16] at 7:15 P.M. central standard time, about 50 people were attending a musical recital in MacVicar Hall located on the Washburn University campus in Topeka, Kansas. Being aware that the area was included in a severe weather forecast they started for the basement when they heard the sirens and the roar of the huge funnel. Someone shouted, "to the southwest corner." In the confusion they sought shelter in the southeast part of the building. They were very fortunate in making this mistake since it saved their lives. The southwest section of the basement was immediately filled with tons of stones and debris by the tornado.

This incident and other somewhat similar cases provided the motivation for this study of the degree of protection afforded by the traditional southwest corner [2,17,18]. The resulting investigation of the houses damaged by the Topeka tornado was conducted in an effort to determine the protection from a tornado offered by particular sections of the basement and first floors of houses without basements.

Data Collection

The tornado funnel (Figure 2-1) that passed through Topeka was about four blocks wide through most of the city. There was almost complete destruction of buildings along the eight-mile path extending from the southwestern edge of Topeka to the city limits on the northeast. The movement of the funnel was from the southwest (Figure 2-2). The tornado has been described as an almost average tornado [19] except for the fact that its path was across a city of 125,000, with a resulting large amount of damage but relatively few deaths. The tornado passed over residential sections and a part of the central business district (Figure 2-3). The funnel was wide enough to encompass the area from the large cylindrical water tower in the foreground to the state capitol building in the background. The large reinforced concrete structures withstood the tornadic winds, although the contents and internal partitions were severely damaged. In the first large building beyond the cylindrical water tower, shown in Figure 2-3 (a 10-story office building that was in the central part of the funnel), all the windows were broken and much of the inside was demolished by the tornado.

27

28

Figure 2-1. Sighting of the Topeka Tornado Caused Immediate Concern for Safety (Photograph courtesy of Topeka Capital-Journal)

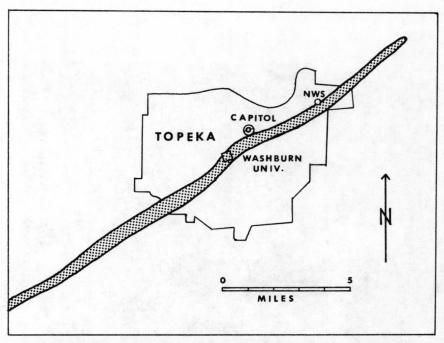

Figure 2-2. Damage Path of the June 8, 1966, Tornado was One-fourth to One-half Mile Wide and 22 Miles in Length as it Crossed Topeka, Kansas, Population 125,000

Within the residential section (Figure 2-4) most of the houses in the tornado path did not have basements and many of the basements under damaged houses were entirely free of debris and, therefore, could not be used in the basement investigation. Since most of the houses in the damage area did not have basements, many people must have sought shelter in some other part of the house. It was estimated that 550 people were injured by this tornado, with about 50 persons injured sufficiently for admission to the hospital. There were only 16 deaths in Topeka—14 of these from injuries during the tornado and two from heart attacks immediately following the storm. It would have been interesting to know the locations of the 550 people when their injuries occurred. However, this information would have been very hard to obtain and would have given less information than other data. Therefore, the investigation was directed toward the inspection of damaged houses in an effort to determine the areas that offered the most and the least protection during this storm.

The data for this investigation were obtained by detailed inspection of individual houses within the damage path of the storm. They fell within

30

Figure 2-3. The Topeka Tornado Crossed the Central Business District Exposing a Variety of Structures to Tornado Winds (Kansas City Star photograph by Sol Studna)

Figure 2-4. Residential Sections Sustained Heavy Damages from the Tornado. About 800 dwellings were destroyed in Topeka. Note the debris in the lower right part of the photograph scattered in the direction the tornado was moving. (Photograph courtesy of Topeka Capital-Journal)

three groups: houses with full basements with one to four feet of the basement wall above ground level; houses with walk-out basements built on a southwest-facing slope so that the southwest wall was almost entirely above ground; and houses without basements whose first floors were inspected. Nearly all of the houses were square or rectangular in shape. Therefore, each basement or first floor was assumed to have nine sections obtained by dividing each outer wall into three equal parts. There were thus eight outer wall sections and one center or middle section of equal area. This did not in general correspond to the room partitions on the first floor that, of course, varied from house to house.

Only houses that had both safe and unsafe areas were used in the investigation. Unsafe areas in the basements and on the first floor of houses without basements were determined by careful inspection. Sections were considered unsafe if they contained piles of boards or glass or chunks of the roof or walls so that a person would probably have been seriously injured if located in that area of the house during the tornado. In some cases it was difficult to determine whether the area would have been safe or unsafe. These areas were marked uncertain during the investigation and were not used in the final analysis.

In the entire damage only 28 full basements were found that had both safe and unsafe sections. The two main reasons for this small number were the low percentages of houses that had basements, and, since only basements with some unsafe area were used, the effectiveness of the basement in furnishing protection. Houses in the southwestern part of the city were all very similar in construction. These were the houses built on the southwest-facing slope that had walk-out basements. These houses were oriented at about a 45° angle with the primary directions. The southwest wall was almost entirely above ground level and the northeast wall was almost entirely submerged. These 17 walk-out basements were treated as a separate group in the analysis.

The investigation of the protection afforded by various areas on the first floor of houses without basements also included only those houses that were severely damaged, yet had some protected area. Therefore, houses that were leveled by the tornado or two-story houses with only roof damage were excluded from the survey. Ninety houses were found to fit these criteria. This number of houses might appear to be a random sample of the estimated 810 dwellings with major damage. However, an effort was made to include all the houses in the damage path that corresponded to these requirements.

The location of each house that was inspected was recorded either by the address or by noting the distance of the house or basement from a particular street corner. The position of the house within the storm path was then determined from a map of the damage path. This was necessary because the damage was about four blocks wide making it very difficult to determine the position of a particular house within the storm track from field observation. Each house was thus determined to be in the northwestern one-third, center one-third, or southeastern one-third of the damage track.

Results and Discussion

Full Basements

The results of the investigation of the 28 full basements are shown in Table 2-1. The largest percentage of the basements was located in the northwestern one-third of the storm path. A Chi-square analysis was performed to determine the dependence of the distribution of unsafe sections on position within the storm track. This gave a value of 5.59 with 16 degrees of freedom that was not significant at the 95 percent level of confidence. Thus, there was no statistical difference in the distribution of unsafe areas in different parts of the storm. The first-floor investigation would be expected to be the most informative with regard to dependence of the occurrence of unsafe sections on position within the storm. However, the results of this investigation, to be discussed later, also showed no significant effect on the distribution of unsafe areas of different locations within the tornado funnel. This is encouraging, since a person interested in protection from an existing funnel would probably not know what part of the funnel was approaching. It should, therefore, be appropriate to consider the relationship between location within the basement and the total number of unsafe sections obtained by summing those in separate thirds of the storm path. Of the total number of unsafe areas, the south-central section of the basement was unsafe twice as frequently as the northeast section of the basement. Other characteristics of the distribution of unsafe areas are shown in Figure 2-5. The respective percentages are plotted for those sections that had the least and the greatest frequency of unsafety.

The tornado carried large quantities of debris (Figure 2-6). It was repeatedly observed during the investigation that the south sides of houses located in the middle and southeastern parts of the damage path were battered by debris carried by the wind. Inspection of many houses in the northwestern part of the storm path did not reveal any similar bombardment of the north sides of the houses. This observation is in agreement with other investigations [20,21] of the direction of damaging winds during a tornado. Damage patterns have shown that trees fell in the general direction of the storm movement on both sides of the center of the damage path.

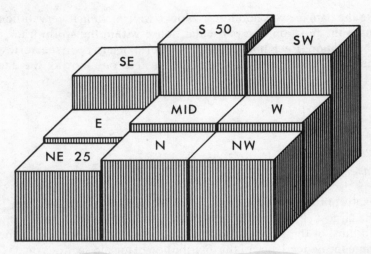

Figure 2-5. Distribution of Unsafe Areas for Basements having both Safe and Unsafe Locations During the Topeka Tornado

Table 2-1
The Frequency of Occurrence of Unsafe Sections of Basements Located in Different Positions within the Storm Path, Topeka Tornado, June 8, 1966

Section of the Basement	Location within the Storm Path			
	NW Third	MID Third	SE Third	Total
SW	8	4	1	13
S	8	5	1	14
SE	6	4	1	11
E	6	2	0	8
NE	5	2	0	7
N	7	2	0	9
NW	7	2	0	9
W	7	2	1	10
MID	7	2	1	10

Exceptions occurred as the tornado funnel lifted off the ground. In this case a convergent pattern of wind direction was indicated toward the point where the tornado lifted. Figure 2-7 shows the amount of debris that battered the south side of an apartment house. The impact was sufficient to cave in some of the above-ground portions of the basement wall. Other basements that had south windows were sometimes unsafe even with no loss of walls because so much debris had blown through the windows.

Figure 2-8 is another example that was repeated several times. The

Figure 2-6. The tornado Carried Debris with it as Shown in This Photograph (Photograph courtesy of Topeka Capital-Journal)

Figure 2-7. Observations Throughout the Damage Path Showed That the South Sides of Dwellings Such as Shown Here were Battered by Debris within the Tornado. Debris and mud did not bombard the north sides of houses in a similar way.

whole house was shifted toward the northeast enough to allow the southwest corner of the house to fall into the basement. When this happened the north part of the basement had much less debris than did the south. This occurred primarily in houses whose foundations were constructed from concrete blocks or stones.

Unsafe sections of the basement other than those exposed to the south were subjected mainly to debris falling through the floor. This seemed to be somewhat of a random occurrence and, therefore, very hard to generalize. Figure 2-9 is included as an example of the way many of the destroyed houses looked. Persons in the basement and away from the south windows would have been completely safe.

The significance of the relationship between frequency of unsafety and location within the basement was evaluated by computing a Chi-square, making the hypothesis that there was equal probability of occurrence of unsafe sections in any location in the basement. The computed Chi-square in this case was 4.15 with eight degrees of freedom. This value was not significant at the 95 percent level of confidence, indicating that the frequencies of unsafe areas shown in Table 2-1 are not significantly different from a random distribution for this number of observations.

Figure 2-8. Houses Sufficiently Moved on Their Foundations so That the Southwest Corner Dropped into the Basement (Photographs by J. R. Eagleman)

Figure 2-9. An Example of a Basement That Offered Protection in all Sections Except Near the South Windows

Walk-out Basements

The results of the investigation of 17 walk-out basements are shown in Table 2-2. These houses were all in the same district and very similar in construction. Each was located on a southwest-facing slope so the southwest wall was almost entirely above ground level and the northeast wall was almost entirely submerged. The southwest side had large windows near each end and two doors near the center. The tornado came from the southwest and almost all of these houses were in the middle of its path. The number of houses in the northwest and southeast thirds of the storm path was insufficient for an analysis of the dependence of distribution of unsafe areas on location within the storm. There is, however, no apparent relationship shown in Table 2-2 and since the other basements and the first floor investigation showed no significant association, the total number of unsafe sections will be considered. The distribution of unsafe sections in the walk-out basements is shown in Figure 2-10. Only 18 percent of the houses were unsafe in the north section, while 88 percent of the houses were unsafe in the south and west sections.

Nearly all of the houses in this group were damaged in the same manner by the tornado. Debris battered the southwest side of the house and came

Table 2-2
The Frequency of Unsafe Areas in the Walk-out Basements Located in Different Positions within the Storm Path, Topeka Tornado, June 8, 1966

Section of the Basement	Location within the Storm Path			
	NW Third	MID Third	SE Third	Total
SW	1	10	0	11
S	2	12	1	15
SE	0	8	0	8
E	0	4	0	4
NE	0	4	0	4
N	0	3	0	3
NW	0	8	0	8
W	2	11	2	15
MID	0	7	0	7

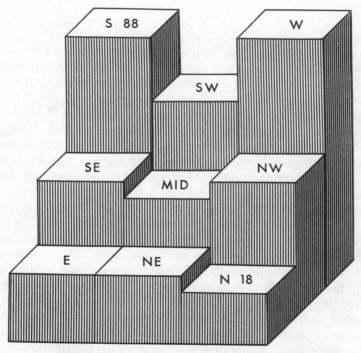

Figure 2-10. The Distribution of Unsafe Areas in Walk-out Basements in Topeka

through the walls or windows into the south and west sections. The ceiling above the southwest rooms of the walk-out basements was blown away in nearly all cases while it frequently remained over the north section. The southwest and northeast sides of two of these houses are shown in Figures 2-11 and 2-12.

The significance of the relationship between location in the walk-out basements and the frequency of unsafety was evaluated by a Chi-square analysis similar to that for the full basement investigation. The assumption of equal probability of safety in all locations within the walk-out basements was made. This gave a Chi-square value of 19.68 with eight degrees of freedom, which was significant at the 95 percent level of confidence. This means that the values shown in Table 2-2 are significantly different from a random distribution and that this distribution of unsafe areas in walk-out basements may have general application. The distribution would not be expected to apply to walk-out basements with a different orientation, however.

First Floors

The frequency of occurrence of unsafe areas on the first floor of the 90 houses used in this analysis is shown in Table 2-3. Eighty-five of these houses were constructed from wood and the remaining five from brick. Each of these houses had both safe and unsafe areas on the first floor. Some of these houses were one-story structures and others were two-story houses that were damaged sufficiently that at least one section on the first floor was unsafe. The stone structures on the Washburn University campus were also investigated but were not used in the analysis since it was felt that they constituted a fourth group. The number of these buildings was insufficient for an evaluation, however.

The effect of location within the storm path upon damage distribution within a given house might be expected to be more important for the first [flo]ors than for the basements. Houses located in the northwest third of the [chan]nel might be expected to have more damage to the north part of the [hou]ses since this should be the windward side because of the cyclonic [circu]lation of the wind. A sufficient number of houses were located in the [north]west part of the damage path so that a Chi-square analysis for the [significa]nce of location within the storm should give good results. The [squ]are value of 4.42, with 16 degrees of freedom, was not significant at [the 95 pe]rcent level of confidence. Therefore, the distribution of unsafe [areas on t]he first floor was not significantly different for separate thirds of [the storm p]ath.

[The distr]ibution of unsafe areas on the first floor is shown in Figure

Figure 2-11. The Southwest Side of a Walk-out Basement in Topeka

Figure 2-12. The Northeast Side of a Walk-out Basement with Ceiling Remaining Over the North Section

Table 2-3
The Frequency of Unsafe Areas on the First Floor of Houses Located in Different Positions within the Storm Path, Topeka Tornado, June 8, 1966

Section of the First Floor	Location within the Storm Path			
	NW Third	MID Third	SE Third	Total
SW	19	25	5	49
S	15	26	5	46
SE	20	28	5	53
E	13	20	3	36
NE	15	20	4	39
N	10	13	4	27
NW	17	18	2	37
W	13	20	3	36
MID	8	16	4	28

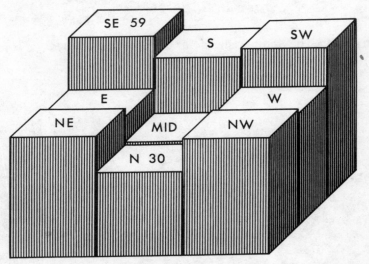

Figure 2-13. The Distribution of Unsafe Locations on the First Floor of Houses in Topeka

2-13. The southeast section was unsafe most frequently and the north-central area was unsafe less often than other sections. A Chi-square analysis of the frequency of unsafe locations, without regard to location within the storm path and assuming an equal probability of safety in all sections of the first floor, gave a value of 16.21 with eight degrees of freedom. This value was significant at the 95 percent level of confidence.

Therefore, an equal probability of safety did not exist for all areas on the first floor. The most unsafe areas on the first floor were those exposed to the south. The southeast corner was unsafe almost twice as frequently as the north-central sections, which was only slightly better than the central area of the house.

Although no records were kept on the protection afforded by rooms of varying sizes it appeared that the smaller rooms were consistently safer. Bathrooms were frequently one of the safe rooms on the first floor especially when they were on the north side of the house or in the central area.

Summary and Conclusions

Results of the inspection of basements under houses severely damaged by the Topeka tornado of June 8, 1966, revealed that for this particular sample the protection offered by any section of the basement along the north wall was considerably greater than the protection near the south wall. However, due to the small sample size this result was not statistically significant.

The investigation of walk-out basements that faced the southwest showed that the north area offered the most protection from the tornado.

The investigation of first floors of houses severely damaged by the tornado was statistically significant and showed that the north and central parts of the house were safer than other locations. If these results are applicable to other tornadoes it is important that the public be advised against seeking shelter from a tornado in the southwest corner on the first floor. Letters from some of those injured in this tornado indicated that this was the location that was chosen in the absence of a basement.

3

Tornado Damage and Safest Locations in Houses

Previous Investigation

Although several thousand persons are injured by tornadoes in the United States each year, increasingly efficient warning systems and improved forecasts mean that people are given more time to seek appropriate shelter from these storms. Most people seek shelter in some part of their house since storm cellars are not very popular in most urban areas. Furthermore, a surprisingly small percentage of houses are being constructed with basements. Therefore, when a tornado strikes a city, most people seek shelter on the first floor of houses without basements. Houses damaged in Topeka, Kansas [16], by a tornado in 1966 showed that the safest location on the first floor was the north part of houses, while the northeast part of basements was safest as described in the previous chapter. Buildings damaged by tornadoes during the spring of 1969 and 1970 were investigated in an effort to gain information on the nature of the winds associated with tornadoes and to determine whether similar patterns of damage occurred to buildings during these different storms.

Observed Damage Patterns

Cities that were struck by tornadoes during 1969 and 1970 were investigated within a few days after the occurrence of the storm. Several general types of structural failures were observed in damaged areas. One common type was extensive roof damage from the weaker storms and also in less damaged areas of the strong storms. An example of this type of building damage is shown in Figure 3-1. Another type of failure that occurred in various regions of the damage path was foundation failure (Figure 3-2). This frequently occurred in houses with concrete block or stone foundations and occasionally with concrete foundations. A third type of damage that was observed was apparent pressure damage to buildings rather than wind damage (Figure 3-3). The fourth type of damage was wall failures in buildings that were apparently quite strong at the foundation (Figure 3-4). Buildings were carefully inspected in order to determine those areas where the probability of injury would have been high during the tornado. At least two persons inspected each house in order to determine from the debris and

45

Figure 3-1. Examples of Roof Damage to Buildings (Photographs by J. R. Eagleman)

Figure 3-2. Examples of Foundation Failures (Photographs by J.R. Eagleman)

Figure 3-3. Examples of Apparent Pressure Damage

Figure 3-4. Examples of Wall Failure

damage whether each of the nine assumed sections of the house should be labeled safe or unsafe. In cases where the evidence from the debris was less definite, the section was labeled uncertain. Houses completely safe or completely unsafe in all areas were not used in the analysis since these provided no information on the relative safety of different parts of the house. The number of damaged houses that had both safe and unsafe areas on the first floor varied from seven in Monroe, Ohio, to 132 in Lubbock, Texas. Very few houses were found in the damage path of these five storms that had basements; therefore, the damage investigation was for the first floor of these houses. It is assumed, however, that a first floor location would be a second choice if a basement were available. Since only houses that had both safe and unsafe rooms on the first floor were used for the investigation aimed at determining the relative safety of various sections of houses, all the buildings illustrated in these figures were not included.

Safest Locations

In addition to the location of unsafe areas, the direction of damaging winds was recorded for houses damaged by the Lubbock, Texas, tornado May 11, 1970. The direction of the damaging winds was determined by such evidence as shown in Figure 3-5. Another type of evidence showing wind directions was the direction of fall of parts of buildings and trees. The location of houses used in the analysis, the direction of damaging winds, and the major damage path are shown in Figure 3-6. The damage path contains two more loops than suggested by another investigation [22]. The center loop explains the major damage to houses on North Cypress, location marked "1" on Figure 3-6. Damaging winds occurred at this location from an easterly direction as well as from a southwesterly direction. The third loop explains the major damage to buildings that would have otherwise been bypassed by the strong winds. These houses were damaged by winds with a northerly component.

Two tornados occurred on May 11, 1970, from the same thunderstorm. All of the houses used in the analysis were from the damage path of the second tornado. Damage from the first funnel located to the east of the later one was minor in comparison, although a large church and some other buildings were damaged by the first funnel. The direction of winds that damaged buildings was predominately from the southwest in the nonlooping part of the tornado damage path. This occurred also in the formation area except that winds from the south and west also damaged some buildings. The number of houses damaged by winds from the various directions is shown in Table 3-1. The numbers in the table represent the number of houses that were unsafe in the various sections during the storm. The

51

Figure 3-5. Indications of the Direction of Damaging Winds During the
Lubbock Tornado (Photographs by J. R. Eagleman)

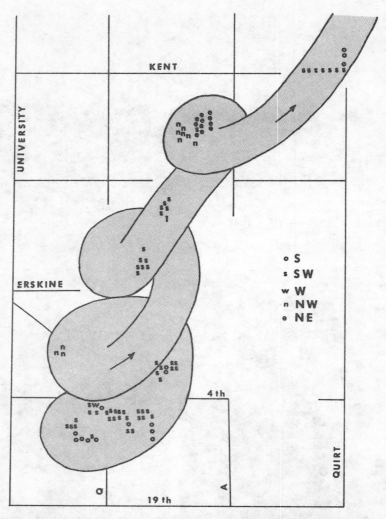

Figure 3-6. Location of Damaged Houses, Major Damage Path, and Direction of Damaging Winds for Lubbock, Texas

largest percentage of the buildings were damaged by southwesterly winds. Eighty-four of the 132 houses or 64 percent of the buildings were damaged from this direction. The effect of the wind direction on the distribution of unsafe areas is apparent from this table. In every case the safest areas were those opposite the direction of the winds. About twice as many houses were unsafe on the windward sides as compared to the areas opposite the direction of the damaging winds. The Chi-square value for houses damaged by southwest winds was improved over that for the total sample indicating

Table 3-1

Occurrences of Unsafe Areas in Houses Damaged by the Tornado in Lubbock, Texas, May 11, 1970

Section of the First Floor	Direction of Damaging Winds					
	SW	S	NW	NE	W	Total
SW	70	20	6	3	2	101
S	69	20	7	4	0	100
SE	65	15	5	7	0	92
E	38	9	8	8	0	63
NE	37	8	10	10	0	65
N	42	6	10	8	0	66
NW	49	10	10	8	2	79
W	63	13	8	5	2	91
MID	42	8	9	6	0	65
Total No. of Houses	84	24	11	11	2	132
Chi-square	28.94	18.07	3.32	6.14	12.00	25.45

a more significant uneven distribution, since the value obtained depends on sample size.

The Chi-square values were calculated by comparison with an even distribution of unsafe areas or the mean number of unsafe areas for each of the wind directions. A Chi-square value of 15.5 is required for significance at the ninty-five percent level of confidence. Previous storms investigated showed no significant influence of location of houses within the storm path.

Damaging winds during the Topeka tornado were much more uniformly from the southwest [16]. As shown in Figure 3-6 the direction of damaging winds was the same as the direction of movement of the funnel for this storm also if the loops in the damage path are considered. Similar characteristics are exhibited by the laboratory model of a tornado vortex to be described in later chapters. The number of loops of a vortex is shown to be related to the speed of movement of the vortex across the surface. Since the Lubbock thunderstorm formed over the city during a period of several hours without much movement, the opportunity was provided for several loops of the funnel.

The distribution of unsafe areas is shown in Table 3-2 for six storms that occurred on May 8, 1969, in Kettering, Ohio; May 10, 1969, in Monroe, Ohio; and June 21, 1969, in Salina, Kansas; April 19, 1970, in Corinth, Mississippi; May 11, 1970, in Lubbock, Texas; and June 8, 1966; in Topeka, Kansas. The tornado approached from the southwest in all the cities except Salina where the direction of movement was from the northwest. It is

Table 3-2
The Number of Occurrences of Unsafe Areas in Houses Damaged by Six Tornadoes

Funnel Movement	SW	SW	SW	SW	SW	SW	NW
Section of the First Floor	Kettering	Monroe	Corinth	Lubbock	Total Four Tornadoes	Total Five Tornadoes Including Topeka	Salina
SW	8	6	18	101	133	182	5
S	10	4	15	100	129	175	6
SE	8	2	17	92	119	172	3
E	5	1	14	63	83	119	4
NE	2	0	11	65	78	117	7
N	3	0	14	66	83	110	7
NW	7	4	12	79	102	139	7
W	5	4	13	91	113	149	7
MID	4	1	7	65	77	105	2
Total No. of Houses	19	7	26	132	184	274	10
Chi-square	14.82	9.62	6.41	25.45	39.48	50.98	5.60

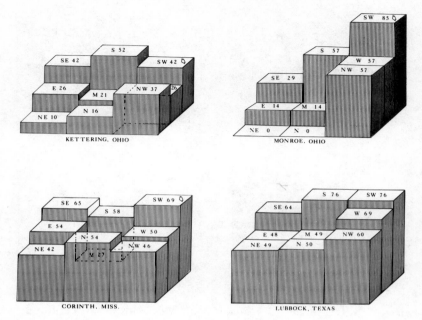

Figure 3-7. Percentages of Houses That Were Unsafe in Various Locations During Tornadoes Coming from a Southwesterly Direction in Ohio, Mississippi, and Texas

apparent that the safest areas in houses during these storms were in rooms opposite the direction of approach of the tornado or located in the middle section of the house.

Figures 3-7 and 3-8 show the percentages of the houses that were unsafe during the storms. These percentages were obtained using the values from Table 3-2 for the number of houses unsafe in each section and the total number of houses investigated after each storm. Besides the Topeka, tornado three of the five cases show that safest areas of the houses were opposite the approach direction of the tornado and the other two show safest areas in the middle of the houses. The Chi-square analysis was performed on the values for the number of unsafe areas against an even distribution or mean of unsafe areas (Table 3-2) for each storm. Probably due to the small sample size, four of the five cases were less then the 15.5 value required for significance at the ninty-five percent level of confidence. The distribution observed for the damage after the Lubbock tornado was significantly different from an even distribution. Safest areas were in the east parts of houses. This was only slightly safer than the middle, north, or northeast parts of houses. This type of distribution of unsafe areas is similar to that obtained after the Topeka tornado for the first floor investigation, which was also based on a large sample and was significant at the ninty-five

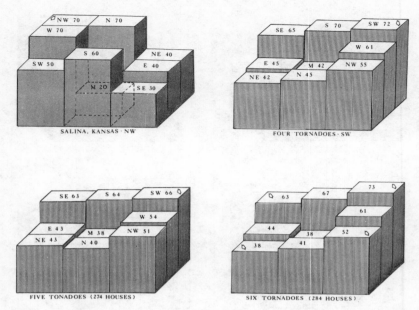

Figure 3-8. Percentages of Houses That Were Unsafe in Various Locations During a Tornado That Came from the Northwest in Salina, Kansas; and Composites of Four Tornadoes That Came from the Southwest; Five Tornadoes That Came from the Southwest During 1966 Through 1970; and Six Tornadoes Combined on the Basis of the Direction of Damaging Winds

percent level of confidence, but showed the north part of houses was only slightly safer than the middle, east, or northeast sections.

Table 3-2 also shows the total number of houses that were unsafe in the various sections for the four storms during 1969 and 1970 that came from the southwest. This included 184 houses investigated. The safest area for this composite group was the middle of the house, with 77 of the 184 houses unsafe in this area. Very simular in safety was the northeast section with 78 of the 184 houses unsafe in this area. Figure 3-8 shows the distribution in percentage values for this composite group. The first floor data from the Topeka tornado in 1966 was also combined with these storms that came from the southwest. This resulted in a sample size of 274 houses and a highly significant Chi-square value. This composite distribution of unsafe areas is also shown in Figure 3-8. These diagrams indicate that the first floor in buildings is considerably safer in the middle, north, east, and northeast parts during tornadoes that approach from the southwest. Even for these five major storms, areas in the northeast, east, north, and in the middle of the buildings were unsafe only 41 percent of the time, on the average, with a range of 38 to 43 percent. Areas exposed to the south averaged unsafe ratings 64 percent of the time, with a range from 63 to 66 percent. It is interesting that the areas exposed to the west were not quite as unsafe as those exposed to the south, with an average value of 57 percent and a range from 51 to 66 percent. This apparently resulted from the greater tendency for the occurrence of damaging winds with a southerly component rather than westerly due to the combined influence from the cyclonic rotation and tornado core flow of these storms that traveled from the southwest.

Figure 3-8 also shows the total sample of 284 houses, considering the orientation of the houses with respect to the damaging winds. The damaging wind direction was the same as the general movement of the tornado funnel, except for Lubbock as shown in Figure 3-6. Seventy-three percent of the houses were unsafe in the windward section, while only 38 percent were unsafe on the downwind side and the middle of the house.

In addition to the relative safety of various locations in houses, the room size was an additional factor. Severely damaged houses that had a variety of room sizes were used to obtain comparative data on this variable. Fifty-seven houses were found to have a sufficient variety of room sizes for comparison. During the investigation, rooms in these houses were ranked from safest to most unsafe on the basis of room size. These data produced a composite ranking from the most safe to least safe of closets, bathrooms, hallways, small rooms, and large rooms. This composite ranking may be somewhat influenced by location since large rooms may be located in outer portions of houses more frequently with hallways and closets located in more interior positions. However, there is some basis for this type of experimental result, since smaller rooms are structurally stronger than larger rooms.

Two additional types of damage were observed. One of these was damage to mobile homes. An example is shown in Figure 3-9 of the extreme damage to these dwellings. House trailers are structurally weak since they are held together in many places by staples rather than nails or bolts. They were frequently destroyed as in this figure that shows only the floor remaining.

Another type of damage is illustrated in Figure 3-10. Automobiles were frequently observed that were demolished as if they had crashed at a very high velocity. These were probably picked up by the tornado and smashed against stationary objects. The object was probably a tree for the automobile shown in Figure 3-10.

No systematic data were obtained on the actual location of persons in houses damaged by the tornadoes since the location of persons injured during the tornado would not provide information on the relative safety of various sections, unless it were assumed that occupants were equally distributed throughout the house. Since this was an unlikely assumption,

Figure 3-9. A Mobile Home Demolished by a Tornado in Hazelhurst, Mississippi (Photograph by J.R. Eagleman)

Figure 3-10. An Automobile Demolished by the Topeka Tornado (Photograph by J.R. Eagleman)

the location of injured or uninjured persons is somewhat meaningless. One particular house was especially interesting, however. Figure 3-11 shows this building that had the closet area in the north central region remaining after the Lubbock tornado. The three occupants of the house were in this closet during the tornado and felt extremely fortunate to be uninjured by the storm. Another interesting aspect of this tornado, also illustrated in this figure, was the condition of the trees. The bark was completely removed from the trees, especially in the region shown in Figure 3-6 that had damaging winds from the northwest. The most severe damage from what appeared to be the most intense winds of any of the storms investigated occurred in this area. It was not readily apparent whether the bark was removed by the abrasive action of particles in the wind or by other forces such as bending and twisting action of the winds.

Storm Direction in Kansas

There is apparently a correlation between the direction of movement of the tornado and the degree of shelter provided by different locations in the houses. Therefore, information on the direction of funnel movement relates to safest locations in houses. The direction of movement of tornadoes across Kansas was obtained for the time period from 1950 through 1970 from *Climatological Data National Summary* [23] for storms prior to 1959, and from *Storm Data* [24] for the rest of the period. The storm paths are shown in Figure 3-12. It is apparent that most of the tornadoes during this time came from the southwest, although some exceptions occurred. Information compiled for the United States by the Weather Bureau [25] shows that 61.4 percent of tornadoes observed from 1930 to 1959 came from a southwesterly direction. The next largest percentage was 16.1 percent from the west, with 10.9 percent from the northwest. These directions accounted for 88.4 percent of the storm directions. Therefore, the previously discussed damage statistics for the five storms that came from the southwest should be representative of the majority of tornadoes, while the statistics from the storm that came from the northwest represents a smaller percentage.

Summary and Conclusions

Six different storms have been investigated for the purpose of determining the relative safety of different locations in buildings and the effect of room size on the relative safety. Four of the six major storms investigated

Figure 3-11. A Closet That Sheltered Three Persons During the Lubbock Tornado (Photograph by J.R. Eagleman)

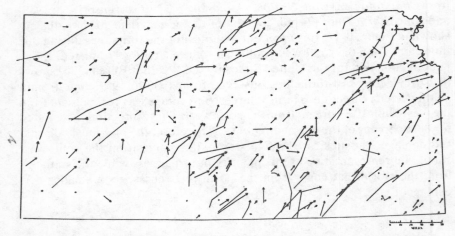

Figure 3-12. Direction and Paths of Tornadoes Occurring in Kansas from 1950-70.

showed the safest location on the first floor of buildings was, in general, the room opposite the approach direction of the tornado. The other two storms investigated showed the central part of buildings to be the safest. The composite sample obtained by combining the data from all five storms that approached from the southwest showed the central part of the houses slightly safer than rooms on the north, northeast, and east. These four sections were much safer than rooms facing the approach direction of the tornado. The total composite sample obtained by considering each house with respect to the direction of damaging winds showed safest areas in the section opposite the damaging winds and in the center of the buildings. These areas were safe about twice as frequently as the sections of buildings facing the damaging winds.

Information obtained on the relative safety of various room sizes resulted in a ranking indicating that the smaller the room size, the greater the safety afforded by the room. The ranking from the most safe to least safe was closets, bathrooms, hallways, small rooms, and large rooms.

4

Theoretical Double Vortex Thunderstorm and Tornado Development Model

Previous Investigations

The flow of air in thunderstorms or mature cumulonimbus has been of interest for some time. Various cumulonimbus models [26] have been proposed by different authors both recently and as early as 1884. One of the problems that has hindered the development of a realistic model has been the lack of direct measurements because of the hazard associated with aircraft penetration of a mature thunderstorm and also the deficiency of suitable instruments for obtaining measurements under such severe conditions.

Radar observations of raindrop distribution in thunderstorms indicate a strong similarity between tornado producing thunderstorms. Similar internal structure is indicated for the supercell thunderstorms since a hook-shaped echo is characteristically associated with them. Another important characteristic of severe thunderstorms is that they are frequently observed to exist in essentially vertical positions in spite of increasing environmental wind shear at higher levels. It has been reported [27] that even small radar echoes change their tilt from the vertical at only 50 to 75 percent of the rate indicated by the ambient wind shear. The cores of convective storms are also reported to be vertical [28] even with extreme wind shear. It was concluded by others [29] that the larger the cloud, the greater will be the relative motion between environment and cloud, since large clouds apparently can withstand a larger wind shear than small clouds. For a nonrotating thunderstorm the vertical drafts and transfer of momentum have been suggested as the source of resistance to the shearing forces aloft. It has been pointed out, however [30], that many large right-moving thunderstorms travel considerably slower than the mean wind. Apparently, the vertical transfer of momentum is not sufficient to explain fully these characteristics of thunderstorms.

It has been suggested [31] that airflow around thunderstorms may be analogous to the flow of fluids around a rotating solid cylinder. Other analyses [32] indicated that the thunderstorm incorporates some environmental air and, therefore, does not completely approximate a solid cylinder. Measurements made from the motion of chaff targets released from an aircraft around a thunderstorm [33] lead to the conclusion that mature thunderstorms do act as effective barriers to air currents in the middle

63

troposphere, while simultaneously drawing some ambient air into the thunderstorm. Doppler wind measurements have also been used [34] to support the assumption that the flow around a mature thunderstorm exhibited essentially the same characteristics as flow around a cylinder. The theory of flow around cylinders has been thoroughly developed from other types of problems and has been applied to the environmental airflow around a thunderstorm [29, 35]. However, these well defined flow characteristics have not been previously used to investigate the corresponding flow patterns that should exist inside a thunderstorm which is a barrier to environmental airflow. Therefore, this chapter deals with the kinematic and dynamic analysis of flow around and within a cylinder that simulates a solid body and describes the thunderstorm and tornado development model resulting from the application of this theory.

Development of the Theoretical Severe Thunderstorm Model

Since the flow patterns around mature thunderstorms are observed to be somewhat similar to the flow around a solid cylinder, an application of the equations describing this flow should give information on possible internal flow patterns of the thunderstorm. The equations [36] for inviscid flow around a solid circular cylinder are

$$V_r = U_o[(A_0/r)^2 - 1] \cos \theta \qquad (4.1)$$

$$V_\theta = U_o[(A_0/r)^2 + 1] \sin \theta \qquad (4.2)$$

where: V_r = the radial velocity.

V_θ = the tangential velocity.

U_o = the environmental free stream velocity.

A_0 = the radius of the cylinder.

r = the radial distance.

θ = the angle.

These equations can be derived mathematically by starting with the source-sink concept or by bringing together two vortices rotating in opposite directions. A solid cylinder, less circular in shape, is also simulated by two vortices separated by a certain distance. In this case the equation for the stream function (ψ) is:

$$\psi = U_o y - (\Gamma/2\pi) \ln (1 + (4sy/x^2 + (y - s)^2)) \qquad (4.3)$$

where Γ is a constant representing the strength of the vortex, and $2s$ is the

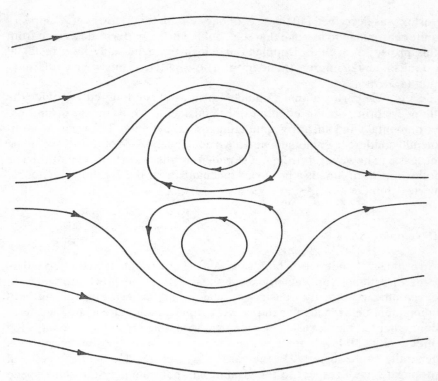

Figure 4-1. Flow Patterns That Would Allow the Atmosphere to Simulate a Solid Body

distance between vortices. Discussions in fluid dynamics center on the flow around the cylinder, but the internal flow is of more interest from a meteorogical standpoint since this could represent the interior of the thunderstorm. Applying Equation (4.3) to the inside of the cylinder gives streamlines as shown in Figure 4-1. It is obvious from Figure 4-1 that this type of flow inside a thunderstorm would provide both the necessary counterbalance flow of air that would oppose the external winds, and a reduction in drag on the outer edges of the cloud that would be most efficient in preserving the shape of the cloud as a vertical cylinder. Individual thunderstorms have been observed to exist for extended periods in areas of strong vertical wind shear. An example in southern Florida is described [37] where a thunderstorm moved very slowly and persisted over a period of several hours although the wind shear in the layer 10,000 to 35,000 feet reached 100 knots.

In addition to Doppler measurements, surface mesolows separated by three or four miles were reported from the Thunderstorm Project [38, 39]. The circulation about the cyclonic vortex was stronger and the anticyclonic vortex was described [40] as a secondary circulation. However, it appears quite reasonable to assume that the double vortex patterns described from this project as well as Doppler radar measurements may be a result of establishment of internal thunderstorm flow patterns consistent with Equation (4.3).

A double vortex within a thunderstorm would undoubtedly provide the thunderstorm with an effective mechanism for withstanding strong environmental wind shear by simulating a solid cylinder. The right environmental conditions combined with a particular sequence of events would be necessary, however, for the development of these vortices in nature. The following discussion is a proposed mechanism for the creation of a double vortex thunderstorm.

Developing Stage

An ordinary thunderstorm builds into a wind environment characterized by large directional and velocity wind shear. The wind profile for tornado development normally consists of winds that have a strong southerly component below 850 mb and that veer rapidly to acquire a large westerly component at mid-levels. The mean wind obtained by averaging the wind directions at all levels from the surface to the top of the thunderstorm is generally southwesterly. In the early stages of development the cell will propagate, with respect to the ground, at approximately the same speed and direction as the mean wind. This movement creates a low level wind relative to the moving thunderstorm with an easterly component, while the mid-level relative winds maintain a westerly component. The winds relative to the moving thunderstorm are thus generally opposing each other between these two levels.

As the thermal updraft carries the low level winds to the mid-levels of the thunderstorm, it collides with the environmental winds. Figure 4-2 illustrates the developing stage of the thunderstorm. The developing cell is moving toward the viewer in this illustration, gathering warm moist air into the front of the lower portion of the cloud. This thermal updraft rises as it crosses inside toward the back of the cloud. As it collides with the opposing environmental winds on the back side of the cloud some of the thermal updraft must rise and double back over the low level indraft and some must split and travel around the sides of the thunderstorm. Therefore, the horizontal cross section through the middle of the cloud begins to approximate a solid cylinder as a double vortex develops, as shown in Figure 4-1, due to the colliding updraft and flow of air around the outer portion at the cloud.

At this early stage the interior of the vortices (round ended cylinders in

Figure 4-2. Double Vortex Flow in the Developing Stage of a Thunderstorm. View toward the west with storm movement from the west-southwest. The vertical kidney shaped regions represent the central cores of the developing double vortex structure.

the illustration) probably contain little or no vertical motion. Kinematically, the horizontal cross section at mid-levels may also include some influence from a source and a sink embedded in a uniform environmental flow, assuming that the slanting updraft expands as it collides with the environmental air providing a source at the back of the cloud. The contribution of a sink in the front part of the storm at mid-levels is probably less than the source since it would arise from more dispersed descending air in the front part of the cloud at mid-levels. The stream function for a source-sink combination in a uniform flow is given by:

$$\psi = U_o y - (K/2\pi) \tan^{-1} (2ys/x^2 + y^2 - s^2) \tag{4.4}$$

where K is a constant related to the strength of the source and the other symbols have the same meaning as in Equation (4.3).

The streamline pattern obtained on an analog field plotter for the flow from a source to a sink embedded in a uniform flow is shown in Figure 4-3. Therefore, there are two possible mechanisms for creation of a double vortex in thunderstorms. Both are related to the thermal updraft. The first arises as a portion of the thermodynamic updraft collides with the external winds, splits to travel around the sides of the thunderstorm, and is strengthened by the viscous shear interaction along the right and left sides of the cloud. These circumstances would develop the flow pattern shown in Figure 4-1. The second mechanism for creation of a double vortex is the effect of expansion of the thermal updraft providing a source at the rear of the cloud. It can be shown mathematically that Equations (4.1) and (4.2) representing flow around a solid cylinder can be obtained from Equation (4.3) by allowing the two vortices to approach each other or from Equation (4.4) by bringing the source and sink closer together. Equations (4.3) and (4.4) represent the flow around a noncircular cylinder that is probably more representative of a thunderstorm.

The streamlines in Figure 4-3 delineate two separate fluid systems with the line corresponding to a stream function value of zero constituting the boundary (ellipse enclosing the source and sink). This line would be equivalent to the surface of a solid body because there is no flow across this boundary line at any location. A similar line exists for the flow around the double vortex in Figure 4-1, although this streamline is not shown. This surface would not necessarily be coincident with the edge of the cloud because it is a function of the relative strength of the separate kinematic components and, therefore, may not correspond exactly to the visible edges of the thunderstorm that represent the liquid or solid particle distribution.

Even in the developing stage of the thunderstorm shown in Figure 4-2, the updraft would provide a hail formation mechanism. Raindrops falling into the strong central updraft would be swept back up between the vortices to the top of the thunderstorm where they would be frozen and unsupported by strong updrafts. They would then fall back into the low level portion of the updraft for a repetition of the cycle. However, the transient nature of the developing stage of the thunderstorm probably prevents the growth of hail stones until a more steady state condition is reached.

As the double vortices develop, the internal flow within the thunderstorm will greatly reduce the wind shear along the edges. Also, if there is a sink at the front of the cell and a source at the back, the turbulent wake will be reduced on the downwind side of the thunderstorm at mid-levels. Either of these factors results in greatly reduced drag on the cell. This reduction in drag allows the thunderstorm to slow down with respect to the ground. A reduction of ground speed results in a reduction in the low level relative wind, since the low level inflow arises primarily from the forward speed of

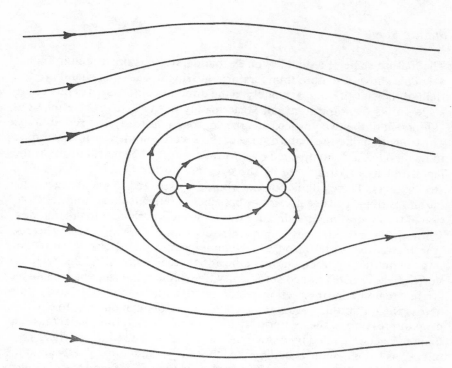

Figure 4-3. Creation of Flow Patterns That Simulate a Solid Body Because of a Source and Sink Located Parallel to the Environmental Airflow

the storm. At the same time, however, a similar reduction in storm speed at mid-levels means an increase in relative winds. Therefore, the thunderstorm may lose some of its capacity to act as a solid body if the internal flow from the updraft is insufficient to counterbalance the external flow at mid-levels. However, any increase in the circulation velocity of the double vortices increases the resistance to the external flow and results in a closer simulation of a solid cylinder.

It has been shown mathematically [41] that a vortex can exist with either an updraft or a downdraft in the center. A core downdraft in the anticyclonic cell resulting from the fall of precipitation would create the same dynamic pressure drop in the center of this vortex as the updraft creates in the cyclonic vortex. Therefore, the thunderstorm could continue to approximate a solid cylinder with a downdraft anticyclonic core and an updraft cyclonic core. The conversion of the core flow in the anticyclonic vortex to the downdraft combined with the penetration of the cyclonic vortex to the jet stream region signifies a transition into the mature stage of the severe thunderstorm.

Mature Stage

The mature stage of a double vortex thunderstorm is illustrated in Figure 4-4. This shows a mature thunderstorm moving toward and slightly to the right of the viewer, approximately in the direction of the anvil. The mature stage varies from the transient stage mainly in the extent of vertical development of the two vortices and the approach to steady state conditions. The transition into the mature stage of the severe thunderstorm, according to this model, is accompanied by several changes. The mechanism for the main transition is triggered as the cyclonic vortex penetrates to the jet stream level. The cyclonic vortex would be expected to be stronger than the anticyclonic cell because of the presence of large-scale cyclonic circulation of the environmental air since this air is incorporated into the thunderstorm through the updraft coming into the front of the storm at low levels. Another reason for the stronger development may be the obstacle effect of large quantities of precipitation in the anticyclonic vortex. As the cyclonic vortex becomes developed sufficiently to extend to the tropopause and jet stream region, the interaction resulting from the unblocked flow of air across the top of this vortex tube would create enough outflow to cause a drop in pressure in the center of the vortex, and, therefore, a decrease in diameter accompanied by an increase in velocity around the vortex. This effect has been demonstrated by the laboratory simulation of a tornado to be described in later chapters. The increased velocity around the cyclonic vortex would reduce the entrainment and decrease the shear on the edges of the storm as the air spiraled around the cyclonic vortex more rapidly requiring more revolutions before getting to the top of the cloud.

Another effect of extending the cyclonic vortex into higher velocity wind fields is the addition of a mechanism for outflow of air through the center of the cyclonic vortex. Therefore, a new updraft region is created as the demand suddenly exists for air through the center of the cyclonic vortex. This updraft can be termed a dynamic updraft in contrast to the other thermodynamic updraft. The dynamic updraft within the cyclonic vortex, Figure 4-5, provides a physical basis for the deduction made by K. A. Browning [42] who examined radar echoes in detail and suggested that a reasonable explanation for the echo-free vault should include an updraft and, perhaps, some circulation. Various investigators [43, 54] have previously mentioned the importance of the jet stream in tornado formation without much suggestion as to the mechanism. It has been verified that a tornado-like vortex can be created in the laboratory by the presence of circulating air combined with suction at the top of the circulating air. The thunderstorm may provide this combination through the dynamic updraft and the double vortex flow of air.

Various energy sources and driving forces probably operate in unison in the mature stage as the thunderstorm approaches steady state conditions.

Figure 4-4. Modified Double Vortex Flow in the Mature Stage of a Severe Thunderstorm. View toward the west with storm movement from the southwest. Vertical cores now contain an updraft in one and a downdraft in the other. Low level inflow at the front of the storm is opposed by the environmental winds in the opposite direction in back of the storm. (Photograph of an oil painting by J.R. Eagleman)

Some of the forces operating in this stage are the unblocked flow of air across the top of the thunderstorm supporting the dynamic updraft core combined with an intensification of the double vortex structure by environmental flow around the outer edges of the thunderstorm. Additional energy is derived from heating during lightning strokes and from the release of latent heat in the thermal updraft surrounding the cyclonic vortex. Thus, the mechanism is provided for a strong downdraft because of falling rain through the anticyclonic vortex that fits the observed characteristics of thunderstorms as well as an intense updraft sufficient to support a tornado. A solid body is simulated by the internal flow patterns of the thunderstorm that allows it to maintain a vertical position.

The thermodynamics are important to the model in several ways. The

72

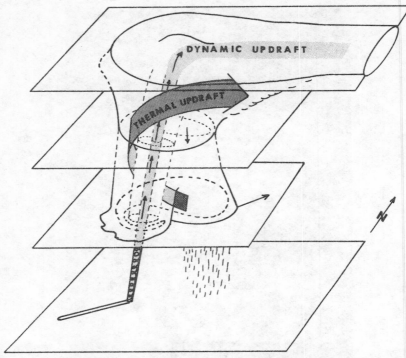

Figure 4-5. Cyclonic Vortex Showing the Distinction Between the Thermal and Dynamic Updrafts. The other major portion of the thermal updraft that rises between the double vortex and around the anticyclonic vortex is not shown.

initial development of the cloud that forms a barrier to environmental flow depends on the proper thermodynamics. Also, the intensity of the thermodynamic updraft in the mature stage is important, since a large inflow of air allows more precipitation to form, which increases the latent heat addition and intensifies the downdraft, reinforcing the flow patterns indicated. The dissipating stage would begin as the thunderstorm moves out of the appropriate wind-shear environment or as the vortices are weakened because of decreased thermodynamic or dynamic support.

As the vortices in the mature stage extend down below the 850 mb level or into the region with relative winds from the front of the thunderstorm, they interact with the weaker low level flow in a different way compared to upper levels. If the strength of the vortices developed from above is sufficient to withstand the tendency of the low level flow to rotate the vortices in the opposite direction, then the main airflow is between the vortices rather than around them as occurs at higher levels. In this case, the lower part of the thunderstorm cannot act as a barrier to flow with patterns

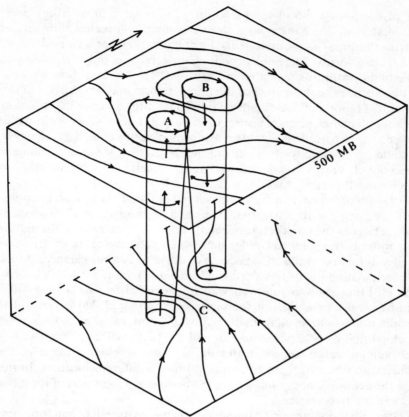

Figure 4-6. Airflow Around a Double Vortex at Mid-levels and Low
Levels with Relative Wind Directions Corresponding to the
Palm Sunday Tornadoes (Flint, Michigan sounding). The positions of the dynamic updraft (A), downdraft (B), and thermal
updraft (C) are also shown.

as in Figure 4-1, but rather sucks the environmental air in between the two
vortices. This low level flow as modeled on the analog field plotter is shown
in Figure 4-6. The lower part of this figure indicates the large quantity of
warm, moist air that would flow into the lower part of the thunderstorm
because the vortices are rotating in the opposite direction for simulation of
a solid body. In addition to the thermodynamic updraft between the vortices, the location of the dynamic updraft is shown in Figure 4-6. The largest
updraft velocity is concentrated in the dynamic updraft in the core of the
cyclonic vortex. However, the thermodynamic updraft contains a larger
volume of air flowing into the front of the thunderstorm at low levels and
provides the double vortex structure. This inflow forms the notch that is

part of the hook echo observed so often on radar returns from severe thunderstorms. The hook echo is caused by a wall of precipatation formed from the thermodynamic updraft and by precipitation that is carried around the cyclonic vortex. The hook echo is characteristic of only the lower part of the thunderstorm as indicated in Figure 4-4. The low level flow between the two vortices near the base of the thunderstorm may also be the source of frequent reports of two clouds colliding just prior to tornado formation. Since the diameter of each vortex is a few miles across, the swirling air between the two vortices would appear as two clouds colliding. However, it should be possible for two cells in the developing stage to combine to form a double vortex thunderstorm. The combination of separate cells has been reported [30].

A family of severe storms was observed [44] that showed an echo-free vault developing at 10,000 to 20,000 feet and reaching the surface about 30 minutes later as the hook echo appeared. The echo-free vault of the mature stage must correspond to the inside of the cyclonic vortex of a mature thunderstorm and would arise because of the large centrifugal force throwing precipatation out of this vortex. This also fits the radar RHI observations [30] that show a southern wall of precipitation and a finger-like, echo-free area extending from the surface up to about 23,000 feet. A wall of precipitation would result from the inflow of warm, moist air at the base of the cloud that rises between and around the two vortices giving rise to a large volume of condensed water drops. The developing stage of the thunderstorm may have an echo-free vault in a different location. In this stage the echo-free area should be in the middle and front part of the cloud between the two vortices.

The mature double vortex thunderstorm has an excellent hail formation mechanism similar to that suggested previously [42] for a different internal thunderstorm structure. Raindrops forming in the thermodynamic updraft would be carried above the freezing level where they would be frozen. Those frozen drops that rise between the vortices would fall back into the low level updraft where they would collect liquid water and repeat the cycle. This would continue until the hailstones grew too large to be supported by the updraft or fell outside the thermodynamic updraft at low levels.

Relative Winds and Thunderstorm Movement

The ideal relative wind profile for the development of the double vortex would consist of winds veering 180° from low levels to mid-levels of the storm. This relative wind profile would exist with winds at mid-levels of the thunderstorm with a westerly component and a forward speed of the

Table 4-1

Measured and Calculated Winds Associated with Severe Thunderstorms, Flint, Michigan, 6:15 P.M., April 11, 1965

Pressure Level	Wind Speed	Wind Direction	Storm Speed	Storm Direction	Relative Wind Speed	Relative Wind Direction
Surface	6 m/sec	160°	17 m/sec	252°	19 m/sec	91°
950 mb	10	161	17	252	20	102
900	11	170	17	252	19	106
850	14	179	17	252	19	117
800	15	202	17	252	14	128
750	18	212	17	252	12	145
700	21	229	17	252	8	175
650	26	232	17	252	11	200
600	27	242	17	252	10	225
550	34	274	17	252	19	294
500	36	250	17	252	19	248
450	46	252	17	252	29	252
400	50	255	17	252	33	257
350	46	255	17	252	29	257
300	45	255	17	252	28	257
250	44	259	17	252	27	264
200	50	262	17	252	33	267
150	36	277	17	252	22	297
100	16	281	17	252	8	6

thunderstorm toward the east sufficient to create low level relative winds with an easterly component. Inflow at low levels is influenced by the storm velocity and will increase if the thunderstorm veers to the right of the mean wind, but would decrease if the storm slows down.

Relative wind profiles were calculated from data obtained from the sounding at Flint, Michigan, on April 11, 1965, at 6:15 P.M., corresponding to the Palm Sunday outbreak of tornadoes and from tornado proximity soundings for various other severe storms. The relative winds in Table 4-1 were calculated by assuming a thunderstorm speed of 60 percent of the mean wind and a direction of 20° to the right of the mean wind direction. This gave a value of 252° for the direction of movement of the Palm Sunday thunderstorm. This compares favorably with values reported for the movement of the radar echoes [45]. The relative winds show a strong shearing environment appropriate for the development of a double vortex thunderstorm. Note in Table 4-1 that the winds relative to a moving thunderstorm are much different from the measured wind speed and direction.

The generation region for the double vortex is probably mid-levels of the storm where the most direct opposition to the low level thermal inflow occurs. Figure 4-7 shows the orientation of the double vortex for four particular storms. These diagrams were constructed from calculations

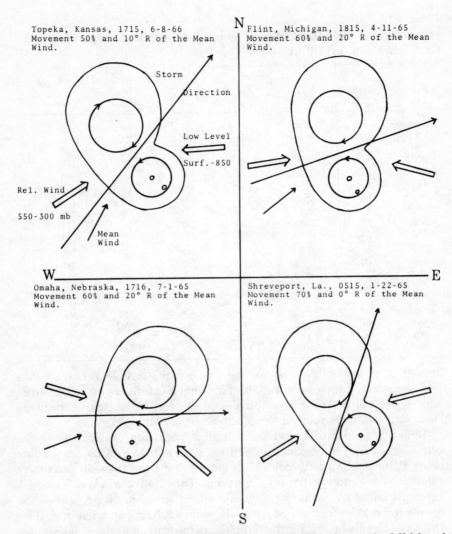

Figure 4-7. Orientation of the Double Vortices in Relation to the Mid-level Relative Winds, Mean Wind, Low Level Relative Winds, and Storm Movement. Calculations are based on tornado proximity soundings.

made from tornado proximity soundings as shown in Table 4-1. The double vortex was assumed to be developed by the winds relative to the moving thunderstorm in the region 550 to 300 mb and is, therefore, oriented with respect to these winds. The mean wind direction computed from an average of wind directions from the surface to 150 mb by 50 mb increments is also shown in Figure 4-7 along with the deviation of the storm direction from the mean wind. The thermal inflow in the front of the thunderstorm provided by the surface to 850 mb level is opposite in direction and of similar magnitude to the mid-level winds.

T. G. Wills [46] concluded that thermodynamic considerations alone were not sufficient to explain tornado genesis and developed a tornado likelihood index based on the strongest shear between any two layers below 550 mb. Various other investigations have reported strong shearing environments during severe thunderstorms, but the double vortex model is the first thunderstorm model that requires this strong shear in order to develop and survive for any length of time.

Observations of the cyclonic vortex in the right rear portion of thunderstorms have frequently been reported. Only during the 1970s, however, Doppler radar observations have been made of an anticyclonic vortex in the left forward part of severe thunderstorms [47, 48, 49]. Such an anticyclonic vortex was observed [47] that first appeared at about 14,000 feet, indicating development at mid-levels of the storm. The obstruction of flow resulting from the presence of large amounts of rain in the downdraft associated with the anticyclonic vortex would cause the anticyclonic vortex flow to be less developed and less frequently observed than the cyclonic vortex.

Some of the variations in the speed of thunderstorms can be explained by considering the amount of drag on a double vortex storm. As the double vortex flow inside the thunderstorm increases, the shear across the outer boundary of the thunderstorm is decreased. This decrease in shear will decrease the turbulent boundary layer along the surface of the cell, which will retard the flow separation and turbulence in the wake of the thunderstorm. Therefore, the thunderstorm can move forward with a velocity corresponding to a lower percentage of the mean wind.

If the cyclonic tube is stronger than the anticyclonic vortex, then the pressure drop on the right side of the cell will be greater than on the left, which will create a lift force pulling the cell to the right. If the upper level wind angle is greater than the storm direction angle, then any movement to the right will increase the upper level relative wind. Likewise, any reduction in the storm speed will cause an increase in upper level relative winds. If the outflow aloft is a function of the relative winds, then a storm that veers to the right and slows down will cause an increase in the outflow at the upper levels. A storm that veers to the right while slowing down would cause the outflow at the upper levels to increase by a greater amount than the inflow at lower levels since the inflow is strongly influenced by the forward speed of the storm. Therefore, various combinations of low level inflow and outflow aloft are possible and are related to the circulation within the thunderstorm, which influences the storm speed and direction.

The pressure pattern frequently observed with thunderstorms is shown schematically in Figure 4-8. The low pressure in the right rear portion of the thunderstorm is associated with the cyclonic vortex and dynamic updraft and the high pressure is associated with the anticyclonic vortex and downdraft core. A second low pressure area is sometimes observed [39, 47]. Apparently, it is associated with the thermal updraft and corresponds to the area beneath the colliding thermal updraft and opposing environmental flow. In addition, this may be a region of new cell development as large surges of the thermal updraft are not likely to always be perfectly matched by the opposing environmental winds resulting in the growth of new cells. It can, therfore, be concluded that the pressure patterns and circulation observed in severe thunderstorms frequently, if not in general, correspond to a double vortex thunderstorm structure.

Comparison of the Double Vortex Model with Other Models

The flow patterns of the double vortex model shown in Figures 4-2 and 4-4 are somewhat similar to the flow suggested by K. A. Browning and F. H. Ludlam [50], especially for a vertical cross section. The horizontal cross section looks quite different, however, and perhaps bears more resemblance to the model proposed by Browning [32]. Flow patterns similar to the Browning and Ludlam model have also been observed by others [51, 52]. These patterns are also consistent with a double vortex thunderstorm structure. Convergence lines and cumulus congestus streets have also been suggested [53] near the location of the lower part of the cyclonic vortex illustrated in Figure 4-4. An updraft tube supporting the tornado has also been proposed [53], but it was assumed that such a tube connected into the thermodynamic updraft of the thunderstorm rather than a second dynamic updraft as suggested by the double vortex thunderstorm model.

The double vortex model provides the physical reason for the occurrence of an updraft of the mature stage as a narrow chimney as suggested by two different models [43, 54].

It was found that in France a very strong wind in the upper troposphere always accompanied severe storms, which led H. I. J. Dessens [43] to believe that the updraft chimney did not draw well without this upper level support. The release of hail at the top of the updraft chimney was also suggested [54]. Both of these models assume a strong slant of the updraft in the downshear direction because of strong wind shear. However, it has been suggested [55] that it is more likely that the main updraft in thunderstorms slants in the upshear direction. The double vortex model offers one of the first logical explanations of the physical mechanism for such an orientation of the thermodynamic updraft.

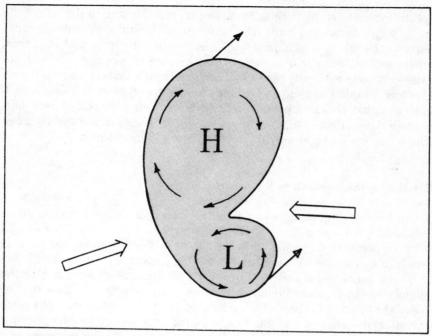

Figure 4-8. Schematic Diagram of Surface Pressure Distributions, Relative Winds, and Storm Movement in Relation to the Double Vortex

Splitting Thunderstorms

The relative wind profile and strength of the thermal updraft are probably quite important in determining whether a thunderstorm will split. There are also some additional forces that may contribute to the splitting of thunderstorms. If the velocity profile in Figure 4-1 is converted to pressure by means of Bernoulli's equation, the areas of relative high and low pressure at the borders of the cylinder can be determined [36]. The areas with forces directed toward the center of the cylinder are on the windward and leeward sides of the cylinder for either the rotating or nonrotating case. Areas of outward forces are concentrated on the edges of the cylinder perpendicular to the relative wind. If viscous rather than ideal flow is assumed, the force directions are similar but the magnitude is somewhat reduced. Therefore, a splitting mechanism is suggested since along with the pressure distribution just described, the magnus effect is in opposite directions for the two vortices and would also tend to separate them. The split of the two vortices would result in cyclonic and anticyclonic rotating cells. Several splitting thunderstorms have been documented [30, 34, 56, 57]. The left cells of splitting thunderstorms are generally observed to move to the left of the mean wind, indicating anticyclonic rotation. Splitting probably occurs early in the life of a thunderstorm, when both cells have updraft cores; otherwise, the anticyclonic cell, or left moving thunderstorm could not exist very long with only the downdraft cell from a mature thunderstorm. Pressure couplets are shown [30] for thunderstorms after splitting, which indicates that the anticyclonic cell develops into a double vortex left-moving thunderstorm that is almost the mirror image of the cyclonic cell which becomes a right moving double vortex thunderstorm.

Relation of the Tornado to the Model

Tornadoes are undoubtedly associated with a larger cyclonic vortex; this is indicated by the pressure patterns observed with tornadoes, showing a drop in pressure from the mesolow, then a more pronounced drop in pressure associated with the funnel. Such a pattern was recorded for the Lubbock tornado [22]. E. M. Brook's [59] analysis of pressure in the vicinity of the tornado funnel showed the surrounding mesolow that he called the tornado cyclone. The collar cloud, which is frequently observed extending from the base of the thunderstorm, probably corresponds to this mesolow and cyclonic vortex as shown in Figure 4-4. The mesolow of a severe thunderstorm is a few miles in diameter and likely extends to the ground for many thunderstorms, resulting in relatively straight damaging winds. This may have occurred during the storm in Lubbock, Texas on May 11, 1970.

The storm developed over the city with very little movement until it reached the mature stage. Winds of hurricane velocity occurred over about a 20 square mile area of the city, as shown in Figure 4-9. The wind direction outside the major tornado damage path, the grey area in Figure 4-9, was investigated from tree fall observations with the location and direction of the damaging winds shown in the figure. It seems quite likely that these high winds were associated with the larger mesocyclone or cyclonic vortex of the thunderstorm. The alternative explanation for this wind pattern is that the sink inside the tornado funnel was of such magnitude to cause this high velocity inflow near the ground. However, since a drop in pressure was measured for the mesocyclone, this pressure sink must have had associated wind velocities at the surface.

The tornado funnel of the double vortex thunderstorm model develops and extends to the ground as the cyclonic vortex of the thunderstorm reaches the jet stream level (Figure 4-10). The cyclonic vortex would be expected to extend upward into the higher velocity winds more readily than the anticyclonic vortex because of dampening of the anticyclonic vortex by

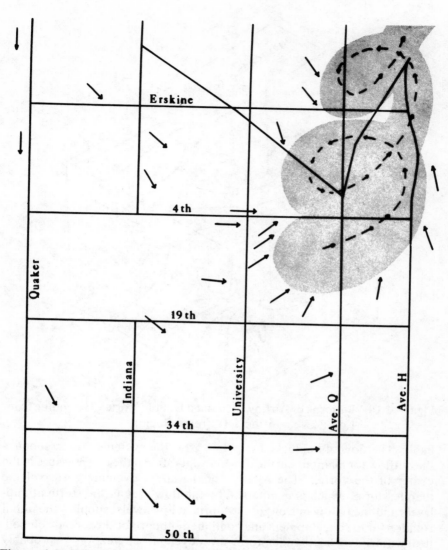

Figure 4-9. Windflow Patterns Outside the Tornado Damage Path in Lubbock, Texas, on May 11, 1970

the falling rain and strengthening of the cyclonic vortex because of a predominance of cyclonic circulation of the larger scale air flow. The intensity of the funnel is probably related to the wind velocity across the top of the cyclonic vortex and the efficiency of the connecting tube supplied by the cyclonic vortex between the higher velocity upper air and the tornado funnel. A combination of several driving forces for the tornado funnel is

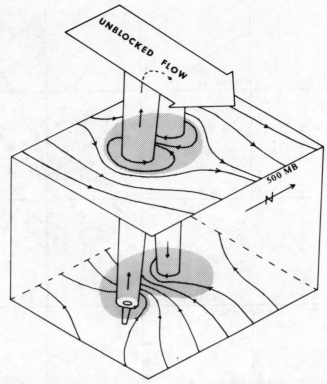

Figure 4-10. Patterns of Airflow Around a Double Vortex Extending from Low Levels to the Jet Stream Region

likely. The flow of unblocked upper air over the cyclonic vortex supplies the outflow for maintaining the dynamic updraft and the low pressure in the center of the vortex. The rotation of the larger cyclonic vortex of the thunderstorm, which is maintained by the flow around the storm at mid-levels and the inflow in front of the storm at low levels, supplies the initial rotation for the development and continued support for the tornado embedded within it. The thunderstorm mesocyclonic circulation is probably sufficient to throw precipitation out of this part of the updraft, which would make the system more efficient through reduced friction and latent energy additions within the mesocyclonic vortex. Since an area of lower pressure is also a favored path of lightning discharge, the dynamic updraft would be an area of increased electrical activity. This would add energy by heating the dynamic updraft core, which would further increase the efficiency of the model. Bernard Vonnegut [60] has calculated the heat input from repeated lightning and feels that it would be sufficient to cause the tornado. There is no doubt that electrical activity increases the energy available for

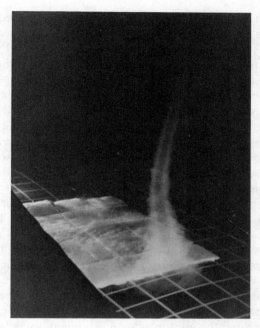

Figure 4-11. An Unconfined Laboratory Vortex Produced by a Combination of Rotation and Upward Core Velocity

the funnel, but this energy probably would not be concentrated until the mesolow associated with the cyclonic vortex became well developed.

It has been verified that an unconfined traveling vortex (Figure 4-11) can be created in the laboratory that has a similar appearance to the tornado funnel. The model vortex is created by a combination of rotation of air and a strong updraft. A thunderstorm that combines a sufficiently strong updraft with some rotation of air would also produce a tornado. Therefore, the double vortex model provides an appropriate mechanism for the development of a tornado by combining the rotation of the cyclonic vortex with the dynamic updraft created by the unblocked flow of air across the top of the vortex.

The shape of the tornado funnel has been reported [40] to be that of a cone when within about two miles of the center of the mesocyclone and rope-shaped when a greater distance away from the center. The tornado funnel frequently forms to the south of the mesolow center [61] or near the center of the cyclonic vortex [40]. These locations are shown by the small circles in the cyclonic vortex in Figure 4-7. Sometimes twin funnels form in separate locations—one near the center or to the right of the mesolow as the first funnel is carried by the mesocyclonic circulation from the right part to the front of the mesocyclone. In several cases of repeating tornado development from a single thunderstorm one funnel originated to the right of the mesolow center and was carried around the mesolow where it dissipated on the left side to allow a new funnel to develop on the right. In the Cleveland storm [40] the funnel probably stabilized near the center of the mesocyclone for a while, allowing it to last much longer than the other individual funnels. The ground path would be a straight line in the same direction as the thunderstorm movement for the funnel in the center of the cyclonic vortex in Figures 4-4, 4-5, and 4-7. The path of the second funnel indicated by the small circle in the right part of the cyclonic vortex in Figure 4-7 would be a curved path as the funnel moved with the circulation of the mesocyclonic vortex and movement of the whole thunderstorm. Such tornadoes forming in this position in the outer part of the cyclonic vortex are periodic tornado producers as the funnels are carried around the front part of the mesocyclonic vortex to dissipate on the left side allowing a new funnel to form on the right of the mesocyclonic vortex. Large severe thunderstorms are frequently periodic tornado producers [62].

Anticyclonic tornadoes apparently occur with a much smaller frequency than cyclonic tornadoes. The theoretical double vortex thunderstorm model would indicate that these tornadoes are associated with left moving thunderstorms that develop from the anticyclonic cell of a split thunderstorm. Given the appropriate environmental winds for additional growth, the anticyclonic vortex would extend to the jet stream region, giving rise to an anticyclonic tornado development mechanism.

These development mechanisms apply primarily to the average and large tornadoes. Small tornadoes sometimes occur on the southwestern flank of thunderstorms and last for a few minutes simply because of opposing winds and shearing action.

Summary and Conclusion

A theoretical double vortex thunderstorm and tornado development model was derived on the basis of kinematic and dynamic analysis of flow inside and around a noncircular cylinder that simulated a solid body. The mechanism for the development of a double vortex thunderstorm was discussed in terms of numerous characteristics of severe thunderstorms that have been measured and observed. New or distinguishing features of the theoretical double vortex thunderstorm model are

1. The inflow into the thunderstorm is provided in the front of the storm at low levels.
2. The thermal inflow at low levels is counterbalanced by the opposing winds at mid-levels and provides the double vortex structure of the thunderstorm.

3. The cyclonic vortex is better developed than the anticyclonic vortex and provides the protection necessary for the development of a second updraft, within the cyclonic vortex.

4. The anticyclonic vortex contains the major part of the precipitation with associated downdraft in the central core.

5. Hail is formed in the thermal updraft while the updraft support for the tornado is provided by the dynamic updraft.

6. A strong shearing environment is required for thunderstorm growth and persistence.

7. The anticyclonic and cyclonic cell of a double vortex thunderstorm may split into separate rotating vortices.

8. Thunderstorm movement is related to the rotation and drag, which is influenced by the double vortex structure.

9. The tornado is related to wind velocities aloft through their support of the dynamic updraft.

10. A source of energy is provided for the tornado from the environmental winds flowing around the sides of the double vortex thunderstorm and over the top of the dynamic updraft.

5 Double Vortex Thunderstorm Characteristics as Simulated on the Analog Computer

Using the assumption that a thunderstorm is a barrier to the wind field as other investigators have done, the application of the equations describing flow patterns around and within such a barrier leads to the development of the double vortex thunderstorm model described in the previous chapter. Since a horizontal slice through the double vortex thunderstorm can be described by mathematical equations, the interactions between this model and the environmental wind field can be analyzed mathematically. More rapid solutions to complicated flow patterns can be obtained by simulations on the analog computer or field plotter [41,63,64]. The operation of an analog computer [65] consists of the use of conductive paper, electrical line inputs, and grounds. The electrical field set up by the various combinations of inputs and grounds can be measured. Therefore, various double vortex flow patterns can be simulated. An error analysis [64] indicated an error within 5 percent of the calculated values with different individuals doing the field plotting. This was considered to be a reasonable error, since very complicated patterns can be simulated on the analog computer and the error is reduced further if one individual does the analysis.

Squall lines are maintained through successive triggering of new convection by lifting of unstable air over a psuedo-cold front, often called "squall front" [29]. Such convection occurring in front of existing vortices could cause one vortex to slow down in relation to the other vortex. New cells are most likely to form within 10 to 15 miles of the upwind end of an existing squall line [66]. The influence of new cells on existing thunderstorms can be determined by analog simulation.

The distinctive character of the severe right moving thunderstorm has been attributed to the intense rotational properties of the updraft [42]. Maximum storm deviation occurs when thunderstorms rotate with a critical tangential speed of a few meters per second [34]. The turning mechanism may be the Kutta-Joukowski force [56]. If a thunderstorm can be considered analogous to a rotating cylinder, as a result of this force, a cyclonically rotating storm would move to the right of the wind field, and an anti-cyclonically rotating storm to the left. If the Kutta-Joukowski force were greater for the cyclonic vortex the decreased friction and pull to the right could cause some reshifting of the double vortex orientation. Several mechanisms that affect the equilibrium of the vortices were discussed in the previous chapter. The upper level winds and the strength of the updraft determine the future of the storm. If the upper level wind backs toward the south, carrying precipitation into the anticyclonic cell, it will have less rotation than the cyclonic cell. The thunderstorm would then be pulled to the right by the magnus force.

Several thunderstorm characteristics are further investigated in this chapter. The first considers cyclonic and anticyclonic vortices that are rotated so that they are nonperpendicular to the environmental airflow. The second deals with cyclonic and anticyclonic vortices separating, but maintaining a position perpendicular to the environmental airflow. Multiple cells and various squall lines are also simulated. The airflow patterns that result when vortices are turned or separated and when they occur in lines are discussed in this chapter.

Vortices Nonperpendicular to Environmental Flow

Since a thunderstorm acts as a barrier to environmental flow, yet incorporates some midtropospheric air into the thunderstorm, it is important to consider the orientation of a double vortex with respect to environmental flow of air. Some indication of the degree of barrier to the air stream can be obtained by determining flow patterns for a double vortex in various positions in a wind field. This has been investigated using the analog computer.

Figure 5-1 shows the airflow patterns as the axis through the vortices is rotated varying amounts in a clockwise manner to the 90° position. One counterclockwise rotation is included to show that flow patterns are similar for a rotation in either direction.

It was shown in the previous chapter that no air flows between the two vortices when they are perpendicular to the air stream. As the axis between the two vortices is rotated 2.5°, no flow results between the vortices using a 5 percent streamline scale (solid lines). However, at a scale of 1 percent intervals (dashed lines), one streamline flows between the vortices. The environmental airflow is starting to shift slightly near the vortices to adjust to the rotation.

For 5° and 10° rotation, one 5 percent streamline flows between the vortices. For these stages, there still appears to be very little change in cell size or airflow around the storm. With 20° right rotation (D in Figure 5-1) two 5 percent streamlines may be observed flowing between the vortices. The inverse of this diagram may also be observed in the 20° left rotation diagram (H in Figure 5-1). With this amount of rotation the size of the vortices is noticeably reduced with considerable air flowing between them. The streamlines are still more densely packed on the outer edges of the vortices than they are on the front or rear.

89

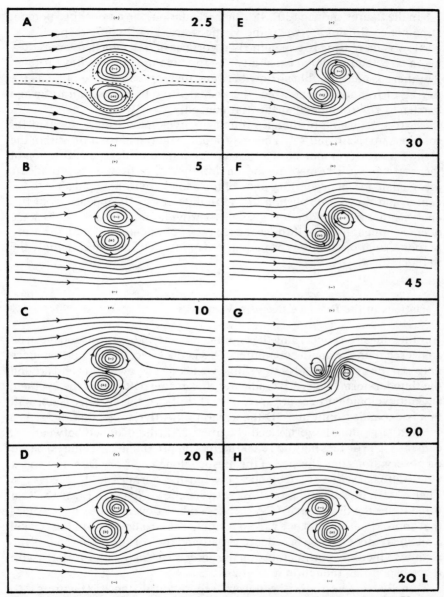

Figure 5-1. Flow Patterns for the Double Vortex Structure as the Vortex Centers are Rotated at Various Angles from the Perpendicular Position. Rotations are shown varying from 2.5° to 90° to the right and at 20° to the left (H), which is a mirrow image of the same degree of rotation (D) to the right.

In the figures showing 30°, 45°, and 90° rotation, the number of 5 percent streamlines flowing between the vortices are 3, 4, and 7, respectively. Airflow is increasing between the two vortices and the vortices are reducing in size and rotational capability. By the time the vortices have rotated 90°, there is considerable airflow between them. It can be seen that as the degree of rotation is increased, the ability of the thunderstorm to form a barrier to environmental airflow is reduced.

Extreme turning of the air around the cyclonic vortex is evident for rotations up to 30° and 45°. This may supply additional rotation within the cloud for the development of a tornado funnel. Some degree of rotation of the vortices with respect to the environmental airflow also provides a mechanism for incorporating some of the cooler environmental air into the center and downdraft portion of the thunderstorm. It appears from this analysis that the thunderstorm is able to provide some degree of blockage of environmental winds through a wide range of orientations with the more perpendicular orientations providing less inflow of tropospheric air into the thunderstorm.

Variations in the Distance Between Vortices

Another question concerning the double vortex thunderstorm model can be answered with the analog computer. This is the effect of various distances between the cyclonic and anticyclonic vortex on airflow within and around the thunderstorm. Figure 5-2 indicates the airflow patterns as the cyclonic and anticyclonic vortices are spread apart. In 5-2A, the distance is normal; this is expanded until in 5-2D the separation is five times that of 5-2A.

In a large double vortex cell (Figure 5-2A), the distance between the center of the vortices may be 10,000 meters (approximately six miles) based on observations that the size of the mesocyclone may vary from two to six miles across. When the distance between the vortices is doubled as in Figure 5-2B, there is still no airflow between the vortices. In Figure 5-2C at three times the distance of 5-2A there is still no flow between the vortices that approximate a solid body. As a result the environmental air must travel a much greater distance in order to get around the vortices. This causes the airflow to increase considerably on the outer edges of the two vortices with a corresponding reduction in pressure.

Finally, at five times the normal separation distance, (Figure 5-2D) there is no dissection at the scale of 5 percent streamlines. However, when Figure 5-2D is observed on a 1 percent streamline scale, three streamlines can be observed flowing between the two vortices indicating that the cell is no longer a complete solid body. These flow patterns for separated vortices are theoretical, since it is difficult to imagine vortices in nature separated by

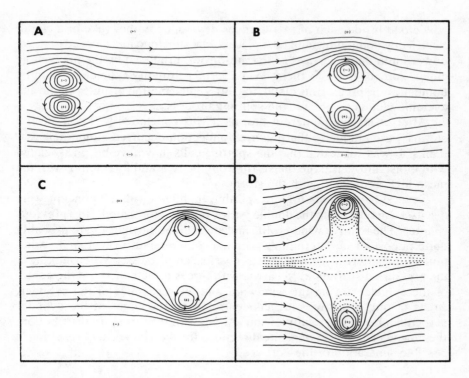

Figure 5-2. Flow Patterns Around a Double Vortex as the Two Vortices are Spread Apart at Increasing Distances

these distances without rotation as in the previous analog computer simulations.

Splitting Thunderstorm Cells

Splitting thunderstorms have been described [30] where the anticyclonic cell develops into a left moving thunderstorm that is almost the mirror image of the cyclonic cell which becomes a right moving thunderstorm. It has already been stated that the Kutta-Joukowski force may result in storms moving either to the left or right of the steering winds [56]. This force may also play an important part in cell splitting. Other explanations for cell splitting include movement to the right of the wind resulting from continuous propagation toward the moisture supply and lower pressure on the right and left edges of the thunderstorm because of flow around the sides.

It has been suggested [34] that splitting involves the development of a new cloud inside each of two vortices developed in the wake of a cloud. Such a development would take place as the subcloud air is forced to move upward into these vortices as the mesofront, produced by the preexisting cloud, spreads beneath them. When further growth of the cloud couplet dominates the preexisting cloud, the differential motion of anticyclonic and cyclonic clouds completes the splitting process.

After observing numerous storms it was concluded [57] that several storms were unique in that intense convection occurred along their rear flanks. It was suggested that the splitting cells may have been associated with moist air overtaking the storm from the rear and interacting with the meso-high associated with the storm.

Several suggestions for the splitting thunderstorm cell concept were discussed in relation to the double vortex thunderstorm model in a previous chapter. The relative wind profile and strength of the updraft are probably quite important in determining whether a thunderstorm will split. Additional forces were also suggested as possible contributors. When looking at the pressure distribution, the areas with forces directed toward the center of the cylinder are on the windward and leeward sides of the cylinder. Areas of outward forces are concentrated on the edges of the cylinder perpendicular to the relative wind. Along with this pressure distribution, the magnus effect is in opposite directions for the two vortices resulting in possible separation of the vortices.

Simulation of Splitting and Nonrecombining Cells

Two variations of splitting thunderstorms are described. The first involves a single cell that splits and eventually forms two cells that in turn split. The remaining vortices (cyclonic to the north and anticyclonic to the south) try to recombine, but are repelled. The second splitting is the same as the first, except when the remaining vortices approach each other. Their paths cross, allowing the anticyclonic vortex to be on the north and the cyclonic vortex to be on the south. In this position it is possible for them to combine into a single cell.

Figure 5-3 describes the sequence of illustrations to follow as Figures 5-4A to 5-4D. Each stage is a separate analog computer plot. The arrows indicate the paths taken by the vortices as they progress from stage to stage.

In the original double vortex thunderstorm (Figure 5-4A) the airflow is equal around both vortices. The splitting of this cell may result in the eventual formation of two cells (Figure 5-4B). In this illustration airflow is more intense in and around the outer vortices than between the two cells.

In Figure 5-4C the two cells split with the two outer vortices dissipating

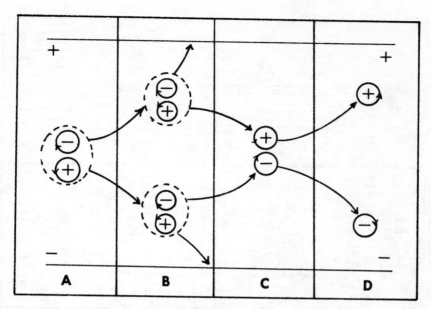

Figure 5-3. Illustration of Splitting Thunderstorm Cells and Positions of the Cyclonic and Anticyclonic Vortices to be Simulated in Figure 5-4 for Noncombining Cells

and the two inner vortices converging. As a result of the two vortices rotating in the opposite direction for simulation of a solid body, the flow is very intense between the two vortices. Little if any rotation is observed at this time. In fact, the vortices would undoubtedly dissipate very rapidly since they are rotating in the opposite direction for simulating a solid body.

In Figure 5-4D the distance between the vortices is double the distance of the previous drawing. Fourteen streamlines flow between the vortices. These diagrams illustrate that it is not possible to form a double vortex thunderstorm from the cyclonic vortex of a split thunderstorm approaching on the north side of an anticyclonic vortex.

Splitting and Recombining Cells

Figure 5-5 illustrates the sequence of illustrations to follow as Figures 5-6A to 5-6F. Each stage is a separate analog field plot, with the arrows indicating the progress from stage to stage. Note that in this sequence, the vortex paths cross between steps C and D.

The first two stages are the same as in the splitting and nonrecombining case. In Figure 5-6C the two cells have split with the outer vortices moving

94

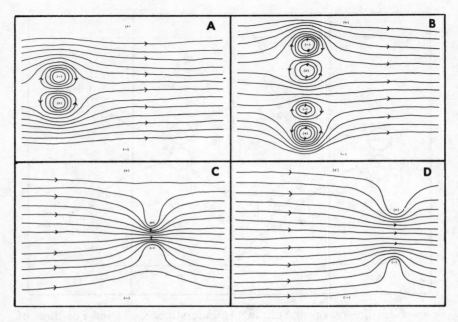

Figure 5-4. Flow Patterns Around Double Vortex Thunderstorms (A and B) and Around Cyclonic and Anticyclonic Cells (C and D) When the Cyclonic Cell is on the Left Hand Side of the Air Stream

off the illustration while the inner vortices converge. The anticyclonic vortex is advancing faster, crossing in front of the cyclonic vortex. This difference in speed would be possible if the cyclonic cell were rotating faster than the anticyclonic cell.

In Figure 5-6D the vortices have moved into different positions, with the cyclonic vortex dropping in behind and to the south of the anticyclonic vortex. There is now some barrier effect and obstruction of surrounding flow. Rotation in the vortices can now be observed with airflow between the vortices reduced. The size of the rotating cells is about equal. The vortices further simulate a solid body in Figures 5-6E and 5-6F as airflow between the vortices is eliminated.

The vortices have recombined and form a solid body with airflow equalized around and within the vortices. Such a state exists because the cyclonic rotation is to the right of the anticyclonic rotation in relation to the environmental wind field, satisfying the internal flow necessary to simulate a solid body. These illustrations show that it is possible for rotating cells to combine into a single unit under the proper conditions. Combination of cells are sometimes observed on radar screens prior to tornado development.

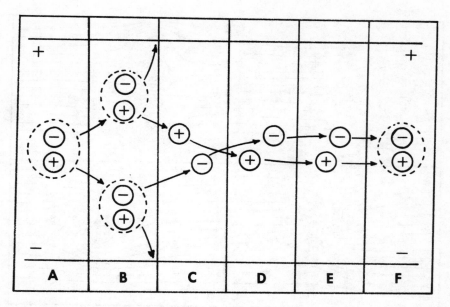

Figure 5-5. Illustration of Splitting Thunderstorm Cells and Positions of
the Cyclonic and Anticyclonic Vortices to be Shown in Figure
5-6 for Combining Cells

Multiple Cell Simulation (Squall Lines)

Characteristics of Squall Lines

This section describes several variations of multiple cells. These variations
include a single straight squall line with a leading cell, a double straight
squall line, and a curving squall line. Observations are made concerning the
effects of given thunderstorm cells on blocking the environmental airflow
between trailing cells.

Squall lines, that is, organized convective systems, are generally associated with thermodynamic instability as well as strong winds aloft.
There are several characteristics that distinguish squall line convection
from ordinary afternoon thunderstorms [29]. Some of the characteristics
are strong asymmetrical character with concentration of convective
phenomena on a favored flank; a tendency for new growth in a favored
direction from the existing thunderstorm mass; the ability of these thunderstorms to persist and even increase during the night hours after ordinary
convection dies out; the pronounced veering of wind with height; and
association with lower and upper level jet streams.

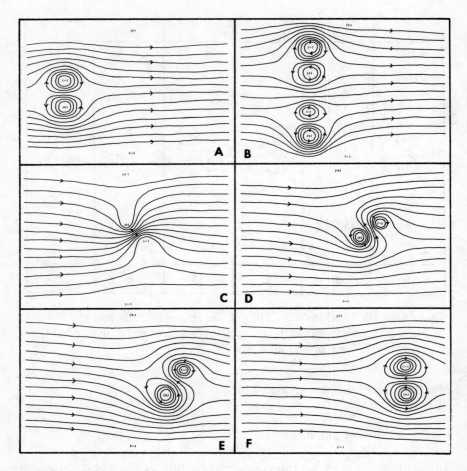

Figure 5-6. Flow Patterns Around Double Vortex Thunderstorms (A and B) and Around Cyclonic and Anticyclonic Rotating Cells (C and D) That Combine into a New Double Vortex Thunderstorm (E and F)

The initial formation of squall lines often occurs near the western edge of the region of potential instability when either hot dry air or the cold front first sweeps into the low level moist tongue. Since such convective systems often last for many hours and extend for hundreds of miles, it was concluded [67] that squall lines are maintained through successive triggering of new convection by lifting of unstable air over a "psuedo-cold front" at the boundary of the rain-cooled air from existing thunderstorms.

With greater stability in lower levels, the energy available for lifting due to spreading out of the downdraft alone may not suffice to set off new convection. Additional energy made available through transfer of kinetic energy from aloft, in the case of strong vertical shear, would make it much more likely that potential instability could be triggered when surface layers are stable. This distinction may account in large part for the fact that squall line type thunderstorms can continue through the night hours when most thunderstorms tend to dissipate [29].

Because squall lines are extensive and can be quite intensive, tornadoes and hail are usually associated with squall lines. Although a single line of echoes usually occurs, in some cases a more complex squall line zone may exist with more than one line. New cells are most likely to form within 10 to 15 miles of the upwind end of an existing squall line at 20 to 40 minute intervals, under conditions of fairly strong winds aloft that veer with height. The cells on the downwind side are generally the oldest ones nearing the dissipation stage [66]. In the case of cell clusters, which are not formed in distinct lines, the most likely region of formation of new echoes is the cloud layer with the largest storms developing on the right hand side of the storm track. Squall lines are not static but continually evolve with individual cells migrating through the storm.

A detailed investigation of the surface weather associated with a severe storm [42] has revealed that the largest hail falls from the intense echo (cyclonic vortex) surrounding the vault, while the heaviest rain falls beneath the left hand parts of the storm. This is consistent with the double vortex model where rain occurs in the anticyclonic downdraft on the left hand side of the storm. In a previous chapter the relation of the tornado to the double vortex model was discussed. It was indicated that tornadoes are associated with a larger cyclonic vortex, this being indicated by the pressure patterns observed with tornadoes showing a drop in pressure from the mesolow, then a more pronounced drop in pressure associated with the funnel. The funnel extends to the ground as the cyclonic vortex of the thunderstorm reaches the jet stream level. The jet stream supplies the outflow for maintaining the low pressure in the center of the vortex. The rotation of the larger cyclonic vortex of the thunderstorm supplies the initial rotation for the development of the tornado.

Results of Squall Line Simulation

Figure 5-7 shows the results of simulations of various types of squall lines. The illustrations are not in a sequence or series, but are merely selected examples of possible patterns that may occur in squall lines.

Figure 5-7A contains three cells in a line with an additional cell approximately 36 miles in front of the line. This cell is parallel with the right hand cell in the line. The purpose of this analysis was to see if there was sufficient

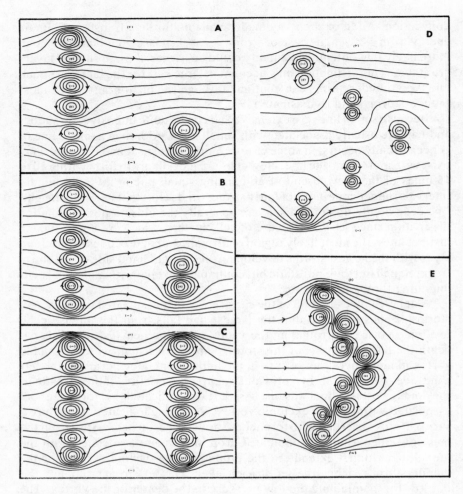

Figure 5-7. Simulation of Flow Patterns Around Multiple Double Vortex
Thunderstorms Arranged in Straight Lines (A, B, and C) and in
Curved Lines (D and E)

airflow through the squall line to support other double vortex thunderstorms in the area. As shown in Figure 5-7A there appears to be adequate
airflow around the thunderstorm that is a short distance downstream from
the squall line. The same conclusion was reached concerning Figure 5-7B,
which also illustrates a double vortex thunderstorm downstream from a
squall line, but located in a different position so that it is in line with the air
flowing between two cells within the squall line.

Figure 5-7C consists of two squall lines, about 36 miles apart, consisting of three thunderstorm cells. It may be observed that airflow is slightly compressed as it passes between the cells; however, the greatest airflow occurs on the outermost edges of the squall line. Given the conditions of Figure 5-7C, if tornadoes were to occur in the southern cell of both squall lines, then a tornado in the second squall line could very likely strike an area that had just been previously struck by a tornado in the first squall line with a time lapse of less than one hour.

Curving squall lines are known to exist in nature. Figures 5-7D and 5-7E model this type of squall line using five cells. In Figure 5-7D the cells are separated by a distance of approximately one and one half times the diameter of a given cell. Airflow is fairly gentle everywhere throughout with the flow being slightly more concentrated on the outer vortices of the two trailing cells and in particular the right rear cell. Nowhere does the solid cylinder concept seem to be hindered. However, the decreased environmental flow would indicate weaker thunderstorms.

In Figure 5-7E the thunderstorm cells are located very close together so that the cells overlap slightly. This simulation indicates that double vortex thunderstorms are possible in a very dense curved squall line. However, airflow patterns are very complex when the cells are located this close together with a considerable interchange of air between separate thunderstorm cells.

Conclusions

The purpose of this chapter was to investigate airflow characteristics associated with a double vortex thunderstorm structure by using analog computer simulation. It was found that complex flow patterns could be adequately simulated. The orientation of the double vortex with respect to the surrounding air stream was modeled. It was shown that the double vortex structure created complete blockage at mid-tropospheric levels if the vortices were oriented so that a line passing through the center of each would be perpendicular to the surrounding air stream. As the orientation of the vortices was changed so that it became less perpendicular to the air stream the amount of inflow into the thunderstorm cell increased proportionately.

The analog simulations showed that if splitting thunderstorms are to recombine, they must be in the proper position, that is, perpendicular to the environmental airflow with the cyclonic vortex on the right and the anticyclonic vortex on the left. This may occur as one thunderstorm cell moves faster than the other one at an angle resulting in a crossover pattern. It was found that vortices cannot form a solid body if the cyclonic vortex is on the left and the anticyclonic vortex is on the right.

Different types of squall lines were modeled. These included straight single lines, straight double lines, and curved lines. The simulation showed that straight squall lines composed of double vortex thunderstorms allow enough of the environmental air stream to flow between the cells to support other squall lines or single thunderstorms located a short distance in advance of the major squall line. It was shown that double vortex thunderstorms can exist in very close proximity, but if the packing is too dense flow patterns become very complex and the cells would undoubtedly be less intense.

6

Theoretical Compressible Flow Tornado Vortex Model

This chapter presents a tornado vortex model based upon observed ground damage and tornado airflow patterns, observed characteristics of an unconfined vortex in free space, vortex theory, and laboratory modeling. Previous investigations such as G.W. Reynolds [21,68] for nine tornadoes, including Vicksburg, Mississippi; Waco, Texas; and Wood River, Illinois, have shown that there were only isolated instances of major damage from winds that were blowing toward the direction from which the tornado came. Reynolds noted that the debris was blown in the direction of translation and that the damage builds up rapidly to a peak and falls off rapidly across the path. He suggested that the speed of translation is an important contribution to the maximum destructive forces. He examined the resulting dynamic pressures across the tornado path as a result of various rotational and translational speeds. The addition of normal translational velocities to the rotational velocities, however, are insufficient to provide the observed damage pattern. He also discussed the static pressures associated with a tornado and noted that chickens are plucked during a tornado. Decompression tests on animals indicate that rapid pressure changes produce no tearing or loosening of the skin. Blast waves and strong winds, however, damage and tear the skin and have plucked chickens.

Investigations were conducted by J.R. Eagleman [16] and Eagleman and V.U. Muirhead [69,70] for six tornadoes, including the Topeka and Lubbock tornadoes. The major damage in each was from winds blowing in the direction of the movement of the tornado tunnel except for local areas in the Lubbock tornado. In these instances an analysis of wind directions and damage indicated that the tornado had made one or more loops. The damage and flow observations were essentially the same as those of Reynolds [21,68]. The wind effect from the vector sum of the rotational and translational velocities was not sufficient to provide an explanation of the observed damage and airflow patterns. The core flow that is very significant must be appropriately added to these two velocities.

If vorticity is present in the air, a vortex with axial core flow will be generated when a mechanism is present to provide a sink for the core flow—the axial core flow organizes the vorticity into the vortex. The characteristics of the tornado are determined chiefly by the axial core flow [71,72,73]. The internal thunderstorm structure required to generate the mesocyclonic vortex and to furnish the necessary blockage of environmental winds was provided by the double vortex thunderstorm model described in a previous chapter. The two vortices are created as the low level thermal updraft collides with opposing environmental winds (Figure 4-4). The mesocyclonic vortex, thus generated, can exist as a steady state vortex for a period. With general cyclonic curvature and thunderstorm rotation, the mesocyclonic vortex is stronger than the anticyclonic vortex. The low level thermal updraft, the double vortices, and the rain provide the partial blockage of environmental air from the low levels to the jet stream region. This blockage varies as the thunderstorm structure changes.

The theoretical compressible flow tornado vortex model provides a mechanism for the triggering of a tornado within the mesocyclonic vortex of the double vortex thunderstorm in its severe steady state stage. An intensity factor is developed from the flow parameters. This factor shows the relationship of these parameters in forming and sustaining the tornado. It provides a means of denoting the severity of the tornado flow conditions. Compressible flow equations are developed for computing pressures, velocites, and densities in the vortex. A local speed index is developed. Equations for axial variations are developed. Surface conditions existing as a result of ground-vortex interactions are discussed.

Tornado Vortex Model

Basic Considerations

The use of incompressible flow equations to describe flow conditions at the high speeds indicated by M.G. Melarango [74] introduces large errors. For example, the dynamic pressure for compressible core flow is

$$q = (P_A - P_O) \frac{1}{I + 1/4M^2 + 1/40M^4 + 1/6000M^6}$$

Thus, for a Mach number less than 0.3, the error in dynamic pressure when the incompressible equation is used is less than 3 percent. As the Mach number is increased to 0.5, the error is 7.5 percent; at a Mach number of 0.8, the error is 17 percent. Since the Mach number increases as the radius of a vortex decreases, large errors will occur in the central region of the vortex by using incompressible flow equations. In the case of the tornado vortex, this is the area of prime interest; therefore, compressible flow equations are required.

The tornado vortex model must satisfy the basic vortex theorems of Lord Kelvin and Hermon von Helmholtz [75] for a three-dimensional vortex. These theorems for a perfect fluid state:

1. Vortex filaments either form closed curves or extend to the fluid boundary.
2. Circulation remains constant throughout the length of the vortex.
3. Circulation remains constant with time.
4. The particles of fluid composing a vortex remain with that vortex.

Although the viscosity of a real fluid modifies these statements for a perfect fluid, the principles must be satisfied by a tornado vortex model.

In the analysis of a vortex it is of prime importance that the existing natural boundary conditions of the flow be satisfied by the theory. Many different boundary conditions have been used in the study of vortices. W.S. Lewellen [76] simulated the flow in a spin rocket motor. In the nonspinning motor the gas flow is entirely outward through the nozzle. By introducing spin the rotational velocity induced a core flow from the atmosphere into the rocket chamber along the centerline of the nozzle (Figure 6-1). When the nozzle flow speed increased to Mach one, the nozzle chocking eliminated the centerline inflow. The spin rocket boundary conditions are in part similar to the hurricane. In both cases the rotational energy is supplied in the region in which the pressure is the greatest and a centerline inflow is produced. In contrast, the tornado rotational energy is furnished at altitudes where the ambient pressure is well below the sea level pressure. Such a change in boundary conditions produces different flow conditions. In addition to the circulation generated aloft there is a divergence aloft, a vertical flow as commonly seen between the ground level and the clouds and a local ground wind condition that produces the damage patterns previously noted.

Model Simplification

In order to simplify the tornado vortex model, the vortex has been divided into vertical and cylindrical regions in accordance with the predominate physical characteristics and/or boundary conditions in these regions. A steady state condition is assumed.

The vertical regions are:

1. *Generation region*: The generation region is the upper region in which circulation and a sink are present to trigger the tornado vortex.
2. *Intermediate region*: The intermediate region of the vortex is that portion of the vortex between the generation region and the ground boundary layer region.
3. *Boundary layer region*: The boundary layer region is the region in which the tornado vortex interacts with the ground. The vortex must

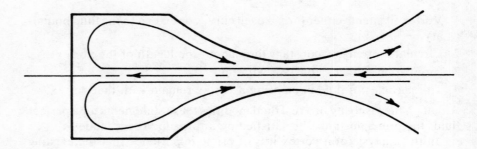

Figure 6-1. Spin Rocket Flow

effectively end in this region such that the basic vortex theorems are satisfied.

The cylindrical regions perpendicular to the vortex centerline are:

1. *Core region*: The core region is that portion of the vortex in which the fluid is assumed to rotate as a solid body. The radial velocity component is zero.

2. *Interaction region*: The interaction region is a thin region between the core flow region and the outer free vortex region in which viscous forces through spiral or Taylor vortices transfer energy from the core flow to the free vortex region. This region is thin and is neglected in pressure calculations in the same manner as the boundary layer region in plane flows.

3. *Free vortex region*: The free vortex region is the region around the core. The flow in this region is predominately tangential and, therefore, the radial and vertical components of velocity are assumed to be zero.

It is assumed that there is no viscous interaction between the free vortex and core flow regions.

Generation Region

To trigger a tornado in the generation region there are three requirements: (1) vorticity present; (2) a sink to establish the core flow that organizes a strong vortex; and (3) conditions that permit the interaction of the two. Sufficient vorticity is apparently present in most severe thunderstorms to supply the circulation considered to be present in the tornado [77]. The second requirement is a sink to establish the core flow. This could be accomplished at the higher altitudes by a relatively low velocity wind relative to the thunderstorm provided it is a cross wind or blowing at a higher velocity than the translation speed of the thunderstorm.

The third requirement suggests that the thunderstorm must be able to block environmental winds up to the sink altitude. If blockage is insufficient, the vorticity is swept downstream and never encounters a sink effect to develop a strong core flow. However, if the blockage enables the vorticity to be carried upward into the sink region and is then swept downstream, a strong core flow is established which organizes the thunderstorm vorticity into a tornado. Figures 6-2 and 6-3 show the overall vortex systems. Eagleman [78] suggested a double vortex thunderstorm structure capable of supporting the tornado vortex.

The mechanism of particular interest in the development of a tornado vortex within the mesocyclonic vortex is the interaction that occurs between the mesocyclonic vortex and the environmental air or jet stream in the upper levels of the double vortex thunderstorm. This interaction depends upon the rate of change of blockage of environmental air flowing toward the mesocyclonic vortex by (1) the thermal updraft within the thunderstorm, (2) the changes in velocity (magnitude and direction) of environmental air with altitude, (3) the position of large volumes of rain inside the thunderstorm, and (4) the position of other cells and updrafts in the surrounding area.

The circulation of the cyclonic and anticyclonic vortices of the thunderstorm is generated mainly in the midlevels where maximum blockage of environmental winds occur. Above the mid-level as these four factors vary, a number of decreasing blockage rates are possible. Three distinct blockage rates are discussed in the following paragraphs.

Small Blockage Decrease Rate. The mesocyclonic vortex, (Figure 6-2) of a severe thunderstorm may have a diameter of the order of 10 miles with a reduced pressure in the mesocyclonic region of several millibars. The maximum vertical air flow Mach number is of the order of 0.1. The core diameter is large and the core pressure at low tangential Mach numbers is nearly equalized radially over the ground surface. Although the mesocyclonic vortex may provide reasonably high surface winds, the maximum Mach numbers are under 0.2 and the flow field can be considered incompressible. The pressure gradients and differentials are also small.

The mesocyclonic vortex must satisfy the vortex theorems. One infinity for ending the vortex must be at ground level and one aloft. Considering the ground level, the viscous torque interaction with the ground is very small because of the large core radius and low tangential velocity. With very little viscous torque developed, the ground forms an effective nonviscous boundary for the vortex and a proper level ending for the mesocyclonic vortex. In the upper atmosphere, the vortex is dissipated over a large distance when the blockage and density are decreasing slowly with altitude. As the core of the mesocyclonic vortex extends upward in the higher levels of the double vortex thunderstorm, it is subjected to slowly increasing effective horizontal wind speeds. A force at right angles to the horizontal wind,

$$\frac{dL}{d\ell} = \rho u_e \Gamma \text{ where } u_e = u_T - u_B \qquad (6.1)$$

curves the core to the right as viewed from the front. Simultaneously a force,

$$\frac{dD}{d\ell} = \frac{\rho}{2} C_D r_m u_e^2 \qquad (6.2)$$

curves the core backward as seen in the side view of Figure 6-3. As these forces are developed on the core, small vortices are shed from the core and trail backward and upward. This occurs over a large length of the core in the upper levels. The core vorticity is slowly dissipated through these small vortices as they stream backward and upward from the curving core. A large area of turbulence is created above and downwind of the thunderstorm. The upper atmosphere provides an effective upper boundary as the pressure and air density decrease and the small viscous damping eventually damps out the small vortices. This upper boundary action of vortex shedding provides sufficient suction force at the top of the mesocyclonic vortex to establish a small updraft within the mesocyclonic vortex.

Moderate Blockage Decrease Rate. When certain critical conditions exist to provide a moderate decreasing rate of blockage of environmental air, the mesocyclonic core vorticity, instead of dissipating over a large distance, is abruptly turned to the right and backward by the effective jet flow. A strong shearing action occurs. The jet trailing vortex (Figure 6-3) is formed within the anvil of the thunderstorm. This new phenomena is nearly one-fifth the diameter of the mesocyclonic vortex (This is analogous to the wing trailing edge vortex [79]). The velocity, pressure, temperature, and density changes associated with the compressible jet trailing vortex are an order of magnitude different from those in the mesocyclonic vortex. The horizontal flow with the imbedded compressible jet trailing vortex provides a large induced drag or low pressure sink near the top of the mesocyclonic vortex, which is an order of magnitude lower than previously existed or would be possible from purely linear flows in the region. A concentrated low pressure region is formed within the mesocyclonic vortex at this altitude. This initiates an intense localized dynamic updraft and triggers the compressible tornado vortex within the mesocyclonic vortex core.

A steady source of energy exists for the tornado vortex as long as the required decrease in blockage with altitude is provided by the thunderstorm to form the compressible jet trailing vortex. The horizontal effective

107

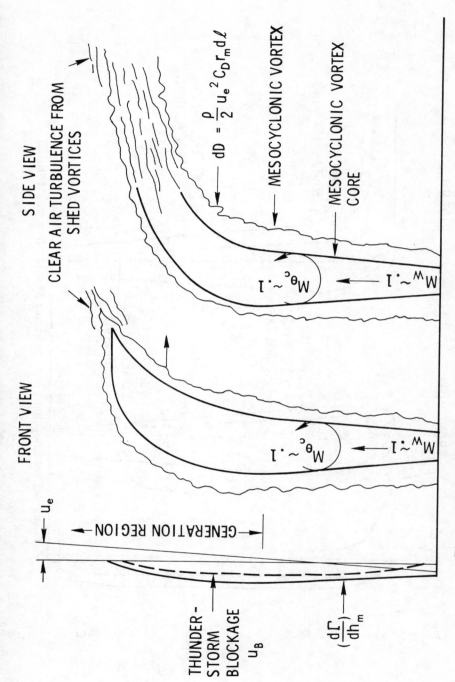

Figure 6-2. Mesocyclonic Vortex During Transient Development Stage

108

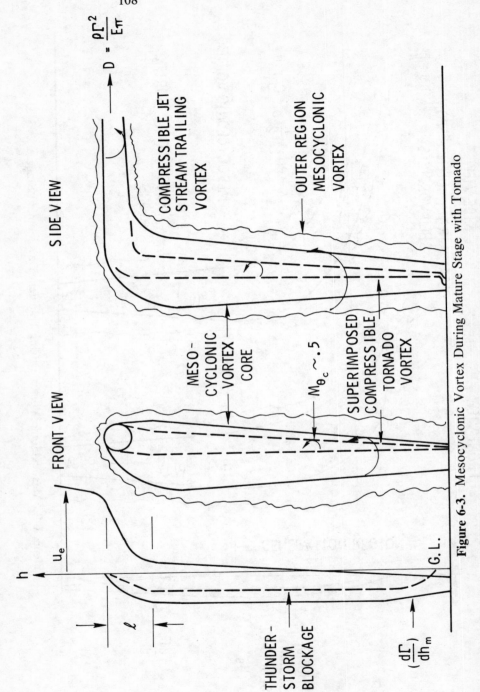

Figure 6-3. Mesocyclonic Vortex During Mature Stage with Tornado

jet stream, in a manner similar to wing trailing vortex, sweeps the vorticity and vertical core flow downstream to an effective positive infinity (Figure 6-4). Thus, the vortex filament created by the shearing action extends to the fluid boundary. The generated vortex filament must also proceed to a minus infinity and seeks the nearest surface, usually the ground. However, in this case the blockage rate change is not sufficient to produce a strong enough funnel to reach the ground.

Using the analogy of the aircraft wing, the circulation in the compressible jet trailing vortex is the same as the bound vortex or in this case, the mesocyclonic vortex. The drag or suction force induced by the compressible jet trailing vortex may be computed as follows:

$$D_i = q_e C_{D_i}(2r_m\ell) \tag{6.3}$$

$$C_{D_i} = \frac{C_L^2}{2E\pi \, \ell/2r_m} \tag{6.4}$$

$$L = u_e\Gamma\ell$$

$$C_L = \frac{\rho u_e \Gamma\ell}{2q_e\ell r_m} = \frac{\rho u_e \Gamma\ell}{\rho u_e^2\ell r_m} = \frac{\Gamma}{u_e r_m}$$

$$C_{D_i} = \left(\frac{\Gamma}{u_e r_m}\right)^2 \frac{1}{2E\pi(\ell/2r_m)} \qquad \frac{\Gamma^2}{E\pi u_e^2 r_m \ell}$$

and

$$D_i = 1/2\rho u_e^2 \frac{\Gamma^2}{E\pi u_e 2r_m\ell} \; 2r_m\ell = \frac{\rho\Gamma^2}{E} \tag{6.5}$$

Thus, when a certain condition is reached to form the compressible jet trailing vortex, a large induced drag force is created that provides a sink pressure over a region less than one-fifth the diameter of the mesocyclonic core (aircraft wing analogy) and an order of magnitude lower than the surrounding region. This induces the vertical core flow of the tornado, which is more than five times that of the original mesocyclonic vortex updraft. The integrated pressure differential over the compressible tornado vortex that is formed would equal D_i.

The conditions aloft required to produce the compressible jet trailing vortex and its characteristic low pressure core are a function of a number of variables (Equations (6.1), (6.2), and (6.5)).

$$L = f(\rho, u_e, \Gamma, \ell) = f(\rho, u_e, v_\theta, r_m, \ell)$$

$$D = f(\rho, u_e, C_D, r_m, \ell) = f(\rho, u_e, \mu, a, r_m, \ell)$$

$$D_i = f(\rho, E, \Gamma) = f(\rho, v_\theta, r_m)$$

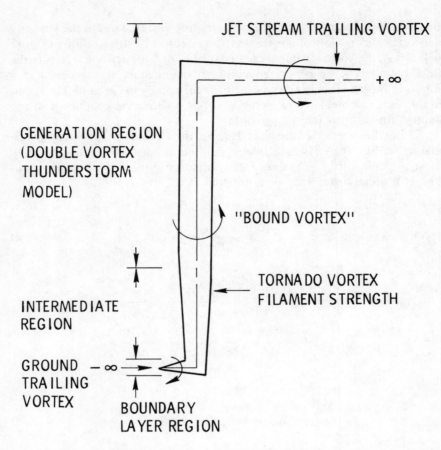

Figure 6-4. Vortex Filament Strength

Defining an intensity factor, $I = 1 - p_o/p_A$, dimensional analysis is used to find the relationship between these variables.

$$1 - I = f(\rho, u_e, [v_\theta r_m], \ell, \mu, a)$$

$$0 = k(M/L^3)^\infty (L/T)^\beta (L^2/T)^\delta \; \delta L^\varepsilon \; (M/LT)^\zeta (L/T)^\lambda$$

$$0 = \alpha + \zeta$$

$$0 = -3\alpha + \beta + 2\delta + \varepsilon - \zeta + \lambda$$

$$0 = -\beta - \delta - \zeta - \lambda$$

$$\alpha = -\zeta$$

$$\beta = \varepsilon - \lambda$$

$$\delta = -\varepsilon - \zeta$$

$$1 - I = k(\rho)^{-\zeta}(u_e)^{\varepsilon-\lambda}(v_\theta r_m)^{-\varepsilon-\zeta}(\ell)^\varepsilon(\mu)^\zeta(a)^\lambda$$

$$= k(u_e\ell/v_\theta r_m)^\varepsilon(\mu/\rho v_\theta r_m)^\zeta(a/u_e)^\lambda$$

$$I = 1 - k(U^\varepsilon/R_\Gamma^\zeta M_e^\lambda) \tag{6.6}$$

where $U = u_e\ell/\Gamma$ (blockage factor)

$R_\Gamma = \rho\Gamma/\mu$ (circulation Reynold's number)

$M_e = u_e/a$ (effective jet stream Mach number)

The intensity factor shows the relationship of the flow parameters in forming and sustaining the tornado vortex. Although the circulation, Γ, is large in the mesocyclonic vortex, I will be nearly zero if u_e is small and ℓ is large. As ℓ decreases, due to a faster decrease of blockage of environmental air with altitude by the thunderstorm, I increases. The strong mesocyclonic vortex, large Γ, will produce a compressible trailing vortex when the proper values of U and M_e exist to produce a critical value of I. The lowest value of I at which the compressible trailing vortex is formed is designated I_1.

Large Blockage Decrease Rate. As the rate of decrease of blockage of environmental air with altitude increases, the value of I increases and the intensity of the compressible trailing vortex increases. The tornado aloft grows downward and reaches the ground. The lowest value of I for this to occur for a given mesocyclonic vortex core radius and altitude of the jet trailing vortex is designated I_2. With values of I slightly above I_2, the tornado vortex at the surface will have a relatively low core flow velocity. As I increases, the core flow velocity would increase.

The distance b that the core extends below the jet stream trailing vortex is a function of both I and the core radius of the mesocyclonic vortex, r_m. If r_m is small, the tornado will be aloft when I is not much larger than I_1 if formed at high altitude. For the same value of I and with a larger r_m a weak tornado may touch down ($b \sim h$). If a weak jet trailing vortex is formed at low altitude with small r_m, a tornado may also reach the ground. As I and r_m increase, larger ground tracks and more damaging surface winds occur ($b > h$). The large damaging tornadoes such as the Lubbock and Topeka tornadoes occur when $I > I_2$ and r_m is very large.

Equation (6.6) confirms the validity of two assumptions made in formulating the compressible vortex theory: First, since the circulation Reynold's number is large in a tornado, the dynamic effects are large compared to the viscous effects in the flow. Viscous effects are only important in thin boundary layer regions. Across these boundary layer regions, the pressure change is negligible. Therefore, the use of a nonvis-

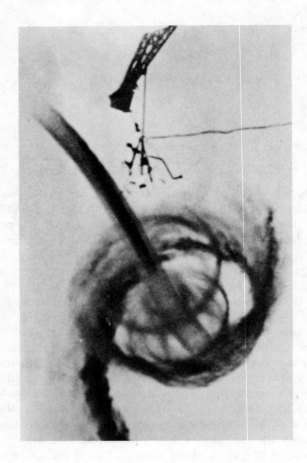

Figure 6-5. Compressible Vortex Formed at the Wing Tip of an Aircraft (Photograph courtesy of National Aeronautics and Space Administration)

cous flow tornado model should yield useful and realistic results. Second, large values of I reflect high flow velocities. Compressibility effects must be considered.

The compressible flow tornado vortex normally would form in the position shown in Figure 6-3 and would be cyclonic in rotation. There is the possibility of the anticyclonic tornado vortex of a splitting thunderstorm triggering an anticyclonic tornado vortex. This would occur when conditions develop in this cell similar to those in the mesocyclonic vortex when a cyclonic tornado is triggered.

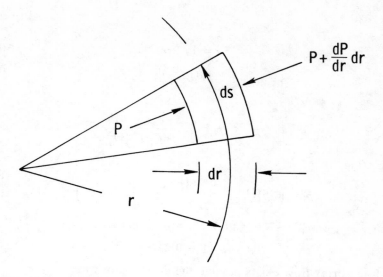

Figure 6-6. Forces Acting on a Fluid Element in Free Vortex Flow

Intermediate Region

As the tornado vortex filament (Figure 6-4) extends toward the ground, the viscosity of the air and increasing density tend to decrease its strength rather than remaining constant as it would in a perfect fluid. However, it is receiving circulation strength from the mesocyclonic vortex in the mid-level generation region. Consequently, the circulation will tend to remain constant. The core mass flow will slowly decrease with decreasing altitude.

Photographs of numerous tornadoes, wing tip (Figure 6-5) and laboratory vortices show two distinct regions of flow. As a result of such evidence, the following assumptions are made in the development of equations for the compressible tornado vortex model:

1. In the core region the radial velocity component (v_r) is negligible.
2. In the free vortex region the vertical (w) and radial (v_r) components of velocity are negligible.

Free Vortex Region Pressure. From Figure 6-6 the force balance for the fluid element in a compressible free vortex is

$$\left(p + \frac{dp}{dr}\,dr \right)\,ds - pds - \frac{\rho v_\theta^2}{r}\,dsr\,dr = 0$$

Since

$$v_\theta = \Gamma/2\pi r \tag{6.7}$$

$$\rho = \rho_A(p/p_A)^{1/\lambda} \tag{6.8}$$

then

$$\int_{p_A}^{p} p^{-1/\gamma}\, dp = \int_{\infty}^{r} \frac{\rho_A \Gamma^2}{4\pi^2 p_A^{1/\gamma}} \frac{dr}{r^3}$$

Integrating:

$$p = \left(p_A^{\gamma-1/\gamma} - \frac{\gamma-1}{\gamma}\frac{1}{8\pi^2}\frac{\rho_A}{p_A^{1/\gamma}}\frac{\Gamma^2}{r^2} \right)^{\gamma/\gamma-1}$$

Using $\gamma = 1.4$, $a_A^2 = \gamma R T_A$ and $M_\theta = v_\theta/a_A$, this gives

$$p = p_A(1 - 0.2M_u^2)^{3.5} \tag{6.9}$$

The free vortex flow exists outside the core region where $r > r_c$.

Core Region Pressure. The pressure within the core may be found as follows:

The core tangential velocity is

$$v_\theta = \omega r \text{ where } \omega = v_{\theta_c}/r_c \tag{6.10}$$

and thus

$$\int_{p_A}^{p} p^{-1/\gamma}\, dp = \int_{\infty}^{rc} \frac{\rho_A \Gamma^2}{4\pi^2 p_A^{1/\gamma}} \frac{dr}{r^3} + \int_{r_c}^{r} \frac{\rho_A \omega^2}{p_A^{1/\gamma}}\, r\, dr$$

Integrating and substituting as before:

$$p = p_A[1 - 0.4M_{\theta_c}^2 + 0.2M_{\theta_c}^2\, (r^2/r_c^2)]^{3.5} \tag{6.11}$$

For the limiting case of a vortex center pressure $P_o = 0$ at $R = 0$, $M_{\theta_c} = 1.58$. For each value of P_o there is a fixed value of M_{θ_c}.

Local Core Flow. With a high altitude low pressure sink and a ground reservoir of air present, a high velocity vertical core flow can be established. Considering the stream line in the core center ($r = 0$ and $v_r = 0$), the flow is vertical. Using the adiabatic energy equation and defining $M_w = w/a_A$, then:

$$w_o^2/2 + C_p T = C_p T_A$$

Since $p = \rho R T$; $C_p/R = \gamma/\gamma-1$ and $a^2 = \gamma p/\rho$

$$p_o = p_A[1 - (\gamma-1/2)M_{w_o}^2]^{\gamma/\gamma-1} \tag{6.12}$$

$$p_o = p_A[1 - 0.2M_{w_o}^2]^{3.5}$$

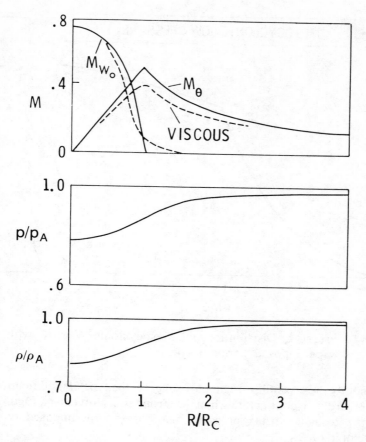

Figure 6-7. Compressible Vortex Mach Number, Pressure and Density Distribution

From Equations (6.11) and (6.12) the maximum vertical flow Mach number at $r = 0$ may be found in terms of M_{θ_c}:

$$p_o/p_A = [1 - 0.2M_{w_o}^2]^{3.5} = [1 - 0.4M_{\theta_c}^2]^{3.5}$$

$$M_{w_o} = 1.414M_{\theta_c} \tag{6.13}$$

between $r = 0$ and $r = r_c$, Equation (6.12) provides the upflow Mach number when modified to account for the tangential velocity component. M_r is zero since v_r was assumed zero.

$$p = p_A[1 - 0.2(M_w^2 + M_\theta^2)]^{3.5} \tag{6.14}$$

The horizontal pressure and density distributions, the tangential and vertical Mach number distributions are plotted in Figure 6-7 for $M_{\theta_c} = 0.5146$. The viscous action between the core region and the free vortex region

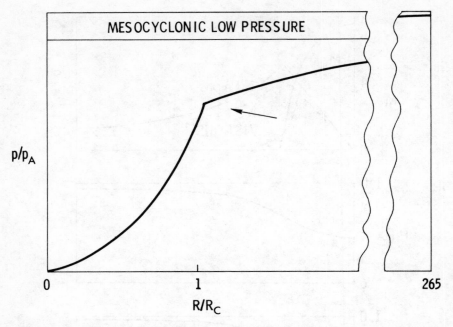

Figure 6-8. Pressure Distribution in Mesocyclonic Vortex with the Superimposed Tornado Vortex

would modify these profiles. Figure 6-7 suggests the type of modification of M_θ and M_w from viscous action when the strong sink aloft exists. Figure 6-8 shows the tornado vortex pressure distribution superimposed on the mesocyclonic low.

Since it has been assumed that the flow around the core is free vortex flow, $v_{r_c} = 0$, the core flow is

$$\mathcal{M} = \int_0^{r_c} 2\pi\rho wr\,dr = \text{constant} \tag{6.15}$$

From Equations (6.11) and (6.14) there is no unique direction of flow or relation between the core diameter and the maximum tangential or vertical velocities. With an infinite reservoir at ground level pressure, p_{A_o}, the mass flow in the core region must be determined by the strength of the sink at the upper level. The sink pressure level and sink area will determine the flow velocities, mass flow, and the resulting values of r_c. The sink pressure level is a function of I, and the sink area is a function of r_m.

Local Speed Index [72]. In order for a core flow to exist in a vortex, a pressure difference must exist along the vortex axis. The minimum core pressure and the maximum axial core flow velocity occur at the core axis (Equation 6.14). The pressure at this point is also a function of the vortex tangential velocities (Equation 6.11). Nondimensionalizing the core axis pressure (p_o) by using the flow reservoir pressure (p_A),

$$p_o/p_A = f(\rho, \Gamma, w_o, \mu, a)$$

From dimensional analysis:

$$p_o/p_A = k/(R_\Gamma^m M_{w_o}^n) \qquad (6.16)$$

where

$$R_\Gamma = \frac{\rho\Gamma}{\mu} = \frac{\Gamma}{\nu} \quad \text{and} \quad M_{w_o} = \frac{w_o}{a_A}$$

A local speed index of the vortex may be formed to indicate the local combination vortex-core strength:

$$S = (1 - p_o/p_A)1{,}000 = \left(1 - \frac{k}{R_\Gamma^m M_{w_o}^n}\right) 1{,}000 \qquad (6.17)$$

As the intensity of the vortex increases, p_o decreases and the speed index increases.

Axial Variations [72,73]. Vortex properties vary along the vortex axis as the speed index varies. Differentiating Equation (6.16) with respect to z:

$$\frac{dp_o}{dz} = p_o \left(\frac{1}{p_A}\frac{dp_A}{dz} + \frac{m}{\nu}\frac{d\nu}{dz} - \frac{m}{\Gamma}\frac{d\Gamma}{dz} - \frac{n}{M_{w_o}}\frac{dM_{w_o}}{dz}\right) \qquad (6.18)$$

From Equation (6.18) and (6.13):

$$\frac{dM_{\theta_c}}{dz} = -0.357\left(\frac{1}{M_{\theta_c}} - 0.4M_{\theta_c}\right)$$

$$\left(\frac{m}{\nu}\frac{d\nu}{dz} - \frac{m}{\Gamma}\frac{d\Gamma}{dz} - \frac{n}{M_{w_o}}\frac{dM_{w_o}}{dz}\right) \qquad (6.19)$$

$$\frac{dr_c}{dz} = r_c\left[\frac{1}{\gamma}\frac{d\gamma}{dz} - \frac{1}{a_A}\frac{da_A}{dz} - \frac{n}{m}\frac{1}{M_{w_o}}\frac{dM_{w_o}}{dz}\right.$$

$$+ \left(\frac{m}{M_{\theta_c}} + \frac{0.3571(1-0.4M_{\theta_c}^2)m}{M_{\theta_c}^2}\right)$$

$$\left.\left(\frac{1}{\gamma}\frac{d\gamma}{dz} - \frac{n}{m}\frac{1}{M_{w_o}}\frac{dM_{w_o}}{dz} - \frac{1}{\Gamma}\frac{d\Gamma}{dz}\right)\right] \qquad (6.20)$$

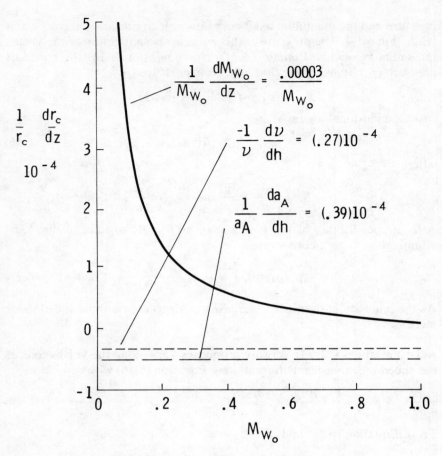

Figure 6-9. Variation of Core Radius

Table 6-1
Variation of Vortex Properties with Height (z)

Change	dp_0/dz	dM_{θ_c}/dz	dr_c/dz
$+dp_A/dz$	+	0	0
$-d\nu/dz$	−	+	−
$+da_A/dz$			−
$-d\Gamma/dz$	+	−	+
$-dM_{\theta_c}/dz$			+
$-dM_{w_0}/dz$	+	−	+

Observations of free air vortices suggest that $d\Gamma/dz$ is a very small value and Γ very large. If we can assume n and m approximately equal and very small, Equation (6.20) can be integrated to give:

$$\ln \frac{r_{c_2}}{r_{c_1}} = \ln \frac{\nu_2}{\nu_1} + \ln \frac{a_{A_1}}{a_{A_2}} + \ln \frac{M_{w_{o1}}}{M_{w_{o2}}} \tag{6.21}$$

The variation of core radius along the axis for each of the factors in Equation (6.21) is plotted in Figure 6-9. The nature of the variations of p_o, M_{θ_c}, and r_c is shown in Table 6-1.

Boundary Layer Region

Ground Action. In a perfect fluid without viscosity, the vortex filament would, if perpendicular to the ground, end at the ground. If a one-dimensional boundary layer could exist, a core inflow might be established as in Figure 6-10. However, these conditions do not exist. The high tangential velocity just above the surface of the ground produces large frictional forces. A torque is exerted on the ground. The reaction to the torque is to roll the vortex in an unstable circular motion. The core is bent along the ground and a ground trailing vortex system resembles the wing horseshoe vortex [79]. The ground viscous action destroys the tangential velocity rapidly. With the core bent along the ground, the core filling from the atmospheric reservoir, p_A, occurs from the direction the core is bent. The inflow velocity decreases as the distance increases from the bend and the inflow region fans out along the surface of the ground. Figure 6-11 illustrates the boundary layer region action. If the intermediate region is stationary with respect to the ground surface, the motion of the vortex on the ground surface is unstable and will whip indiscriminately over the surface in erratic circular patterns. The strong core inflow acts like a vacuum sweeper hose. The longer the intermediate region remains stationary over an area, the larger the affected ground area and the more complete the destruction from this erratic motion.

When the intermediate region is moving relative to the ground surface, the viscous action between the ground and the ground trailing vortex produces a drag on the ground trailing vortex, Equations (6.1), (6.2), and (6.5). This drag has a stabilizing influence on the motion of the core relative to the ground. If the translation speed of the intermediate region is small, a small drag results and the resulting ground trace is a series of connecting loops or cycloidal marks (Figure 3-6). The spacing and size of the loops are a function of the translation speed, the core diameter, and the maximum core tangential velocity. As the relative motion over the ground increases,

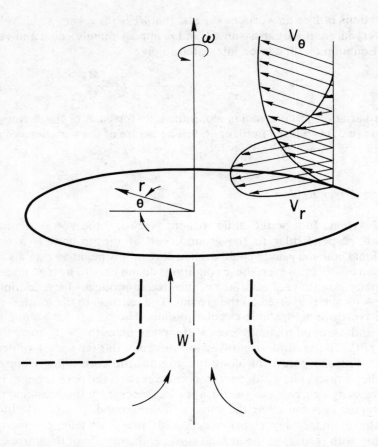

Figure 6-10. Ideal Core Filling in Boundary Layer Region with Core Perpendicular to Surface and One-dimensional Viscous Interaction

the circles become fewer and eventually transform into a wavy pattern. At a still higher speed the track becomes nearly straight.

Instantaneous Velocities. The instantaneous ground flow velocities at any point are the vector sum of the tangential velocity, the core inflow velocity, and the core translation velocity. Local instantaneous velocities and streamlines are illustrated in Figures 6-12 and 6-13 for a translation speed that gives a nearly straight track. It can be seen that the major damaging flow velocity is approximately in the direction of the translation velocity of the core. Where the high velocities exist the local pressure must be low. Therefore, not only should buildings appear to be damaged by a high

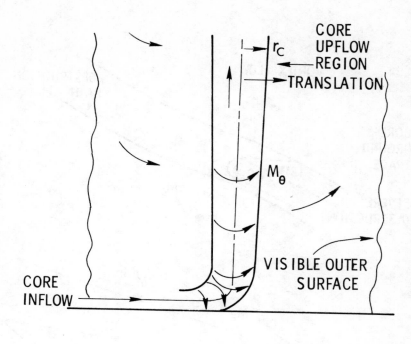

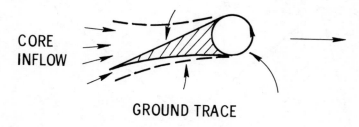

GROUND TRACE

Figure 6-11. Core Filling in Boundary Layer with Vortex Surface Viscous Interaction

straight wind, but also the low resulting pressure will leave suction marks on soft soil and other viscous materials. Equations (6.5) and (6.17) are applicable to the ground trailing vortex core flow. Figure 6-14 shows the instantaneous resultant velocities at the surface perpendicular to the translation axis.

Destructive Effect. The destructive effect of the winds is shown in Figure 6-15 by a plot of the dynamic pressures:

$$q = (\gamma/2)pM^2 \qquad (6.22)$$

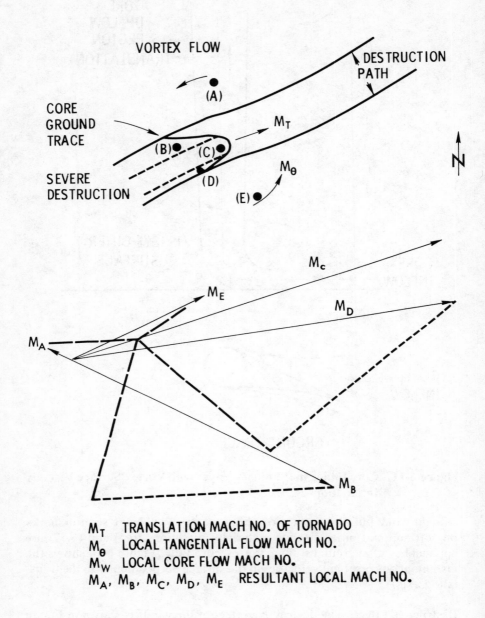

Figure 6-12. Instantaneous Local Ground Flow

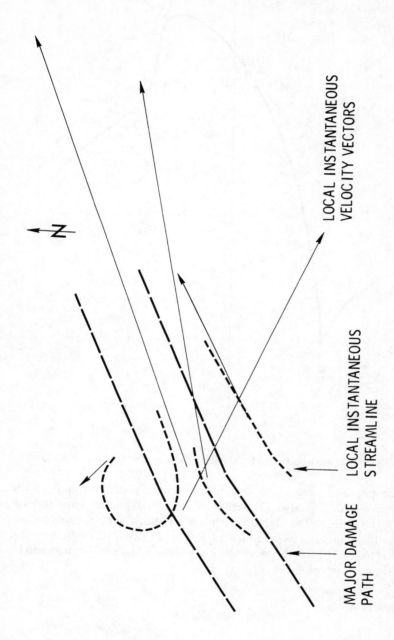

N

LOCAL INSTANTANEOUS
VELOCITY VECTORS

LOCAL INSTANTANEOUS
STREAMLINE

MAJOR DAMAGE
PATH

Figure 6-13. Instantaneous Local Ground Flow Streamlines and Velocities

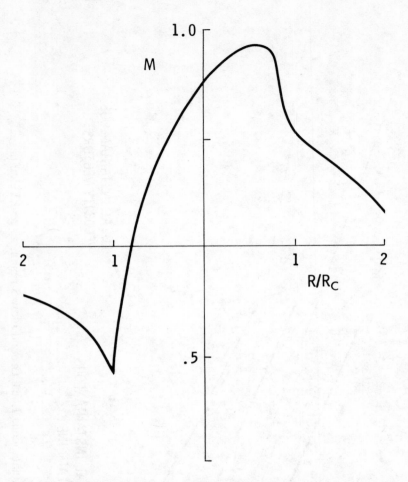

Figure 6-14. Instantaneous Surface Mach Numbers (Cyclonic Rotation)

The force per unit area on a structure is directly proportional to the dynamic pressure. It is also a function of the aerodynamic shape (coefficients) of the object. The dynamic pressure pattern agrees qualitatively with both the damage observations of Reynolds [21,68], Eagleman and Muirhead [16,69,70] and the laboratory simulation of a moving tornado by Muirhead and Eagleman [71].

Illustration of Model

Ground Conditions

The path of the Lubbock tornado—second—(Figure 3-6) was developed

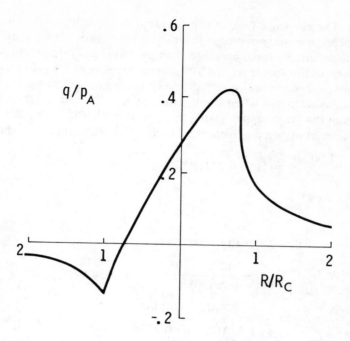

Figure 6-15. Surface Dynamic Pressure with Vortex Translating (Cyclonic Rotation)

from observed wind directions and the boundary layer flow theory of the previous section. The path was also developed from observed wind directions and observed damage by Eagleman and Muirhead [69]. The two paths were nearly identical. In investigating the tornado damage to several hundred structures at Lubbock, the damage appears to be from a strong straight wind. Investigation of the damage caused in other tornadoes during the past four years has also indicated that the damage was caused from a strong straight wind. It was observed after the Topeka tornado that a considerable amount of debris accumulated on the side of the houses facing the approach direction of the tornado, according to Eagleman [16,69]. Numerous houses with concrete block or stone foundations were observed to be moved a short distance in the direction of the movement of the funnel. The boundary layer flow of the theoretical compressible flow tornado model provides an explanation of these observations.

The Lubbock Weather Bureau pressure trace was used with the previously developed equations to compute the core conditions. The Weather Bureau Office was located in the path of the core; however, the damage in the area did not indicate that the full effect of the core was felt on the ground. Either the tornado had lost intensity or was moving very rapidly at this point. The response time of the standard barograph together with its internal location would not enable the instrument to provide the sensitivity required to record the rapid pressure changes in the core region. However, the trace shows the mesocyclonic low pressure region with a portion of the superimposed tornado pressure region. Recording of the pressure should be reasonably accurate when the instrument was located as far as 1,610 meters from the core center. Using this pressure, 1,000.2 mb, [22] and a core diameter of 610 meters from the damage path, Equation (6.9) shows:

$$\frac{p}{p_A} = \frac{1,002.0}{1,008.7} = [1 - 0.2M_\theta^2]^{3.5}$$

$$M_\theta = 0.097$$

and

$$\Gamma = 2\pi(1610)(0.097)(344) = 10.72\pi \cdot 10^4 \tag{6.7}$$

$$V_{\theta_c} = \frac{10.72\pi \cdot 10^4}{2\pi(305)} = 177.0 \text{ m/s} \tag{6.7}$$

$$M_{\theta_c} = 0.514$$

$$M_{wmax} = 0.7278 \tag{6.13}$$

The resulting Mach number distribution of vertical core flow and tangential velocity, the density and the pressure vs. r/r_c are shown in Figure 6-7. The corresponding instantaneous surface Mach numbers and dynamic pressure across the Lubbock tornado path are given in Figures 6-14 and 6-15. From the plot of dynamic pressure as a function of radius, the devastating nature of the core flow can be readily seen. The aerodynamic force on an object in subsonic compressible flow is

$$F = \frac{qC_FQ}{\sqrt{1 - M^2}} \tag{6.23}$$

C_FQ is a fixed value for a given object and wind orientation. Not only is q much larger in the core region than in the tangential flow regions, but, also, the $\sqrt{1 - M^2}$ term is much less. Obviously, Equation (6.23) is not valid in the region near Mach one and must be replaced by a more exact expression from the nonlinear differential flow equations governing the transonic flow regime [80].

Computed Conditions Aloft

Photographs of tornado vortices such as Figure 6-16 show a slowly decreas-

Figure 6-16. Ellsworth Tornado (Photograph courtesy of the Kansas
Highway Department)

ing core radius with decreasing altitude below the generation region. Using
a value of dr_c/dz from photographs together with Equation (6.21) and
standard atmospheric values, Table 6-2 was constructed from the Lubbock
surface data. The core radius is plotted in Figure 6-17.

Simulation of Theoretical Model

Boeing 737 Engine Inlet Investigation [81]

During the normal operation of the Boeing 737 over gravel runways,

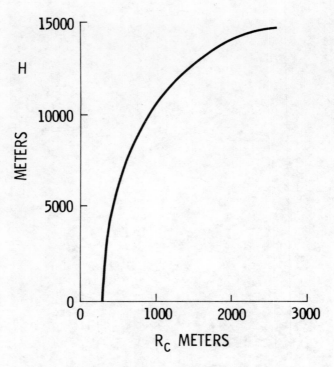

Figure 6-17. Variation of Core Radius

excessive foreign object damage was sustained when cross winds were encountered on gravel runways. Subsequent investigations disclosed that a tornado-like vortex was developed by the cross wind impinging on the air being pulled into the engine inlet. A jet engine test installation was set up to produce the small tornado vortex. Extensive tests were conducted to determine an optimum installation on the 737 to preclude the vortex formation. The test setup is shown in Figure 6-18. It will be noted that the jet engine provides a strong, horizontal airflow above the ground. Rotation and an intersecting side draft were provided by fans and the propeller. From the combination an intensive vortex was established ahead of the engine inlet. The upper end of the vortex terminated in the engine inlet. The other end sought the nearest flat surface, usually the ground. Occasionally, a second vortex appeared, one going to the ground, the other to the fuselage.

The vortex was intense enough to draw rocks and other foreign material into the intake. In this instance the large horizontal flow into the jet provides the mechanism to establish and drive the intense vortex.

The Boeing 737 engine inlet problem closely simulates the tornado

Figure 6-18. Boeing Aircraft Company Jet Engine Vortex Generator (Photograph courtesy of the Boeing Company)

generation conditions aloft. Photographs of the vortex were made by the use of dry ice and water on the ground surface.

University of Kansas Laboratory Model [71]

A reasonable simulation of Mach number and boundary conditions of the theoretical compressible tornado model was provided by the laboratory vortex generator (Figure 6-19). Cage rotation provided circulation at a height well above ground level. A small hole in the center of the top of the cage provided a variable outflow. The outflow was controlled by varying the pressure at the hole. Dry ice was used for visualization. The values of Mach number, pressure, and density were computed from the measured minimum core pressure ($r = 0$) and plotted in Figure 7-7. Measured model pressure data at the other radii are also shown. The ground trailing vortex core track data (Figure 6-20) also substantiate the theoretical compressible flow tornado vortex model.

Conclusions

1. The tornado vortex model with strong core flow provides an expla-

Figure 6-19. University of Kansas Tornado Vortex Simulation

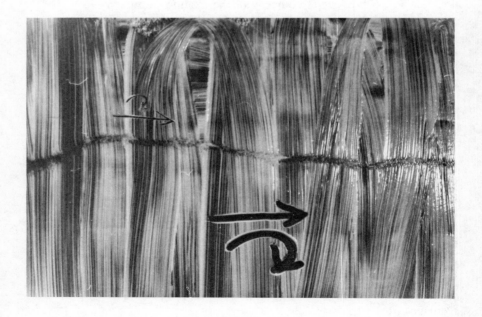

Figure 6-20. Ground Trailing Vortex Core Track and Flow Pattern

Table 6-2
Variation of Core Radius with Height

h Meters	r_c Meters	M_{w_0}
0	305.0	0.7278
3,000	355.4	0.8542
6,000	405.8	1.0420
9,000	711.5	0.8542
12,000	1267.0	0.7278
15,000	2451.1	0.6014

nation of the observed localized nature of the tornado, tornado damage, tornado paths, and ground flow after tornado passage. Laboratory modeling of the flow model produced the same patterns.

2. The theoretical compressible flow tornado model, which was presented, satisfies the basic vortex theorem of Lord Kelvin and Hermon von Helmholtz and the existing natural boundary conditions.

3. The model provides a mechanism for the establishment of a strong compressible flow vortex superimposed in the mesocyclonic vortex that is an order of magnitude greater than the mesocyclonic vortex. The combined double vortex thunderstorm model and the theoretical compressible flow tornado model provide an overall theory for tornado development and the sustaining of the tornado for a finite time period. An intensity factor that was developed relates the vorticity and blockage factors to determine tornado vortex intensity.

4. The speed index (Equation 6.17) derived by dimensional analysis provides a measure of the intensity of the tornado vortex at any point along the axis. Conditions along the vortex axis are obtained by integrating Equations (6.18) and (6.21). However, a systematic flight measurement program is needed to determine the value of the exponents (Equation 6.17) and the rates of decay of the core flow and circulation in large-scale vortices. Such a program would also enable the determination of the variations across the vortex from the isentropic conditions assumed in Equations (6.7) through (6.15).

5. Although the calculations made on the Lubbock tornado (Table 6-2) contain a number of assumptions and are not reliable quantitatively, the calculations illustrate qualitatively the phenomena that exists in a double vortex thunderstorm when I is greater than I_2.

7

Simulation of a Moving Tornado in the Laboratory

The prototype tornado modeled in the laboratory is shown in Figure 7-1. This theoretical compressible flow tornado vortex was developed together with the supporting double vortex thunderstorm and tornado development model presented in previous chapters. In order to simplify the theory of the tornado vortex, the vortex was divided into vertical and cylindrical regions in accordance with the predominate physical characteristics and/or boundary conditions.

Vertical Regions

Generation Region. As a mature severe thunderstorm develops, vertical rising air currents with rotation are produced. The mesocyclonic vortex of a large, severe thunderstorm may have a diameter of the order of six miles with reduced pressure in the mesocyclonic region of several millibars. Although the mesocyclonic vortex may produce high surface winds, the velocities are well below those in the tornado vortex. The vertical air flow velocity is also relatively small.

When a strong mesocyclonic vortex develops within a double vortex thunderstorm and extends upward to the region of the tropopause and encounters a strong horizontal flow or jet stream, a tornado is triggered within the mesocyclonic vortex by the large amount of additional energy available from the horizontal air stream. As the rotation air current or the mesocyclonic vortex encounters the jet stream, a strong shearing action is set up and a jet trailing vortex is formed. This action is similar to the formation of a wing trailing vortex on an aircraft [79]. The horizontal flow of the jet stream with the embedded trailing vortex provides an induced drag force or low pressure sink for the establishment of a strong localized vertical core flow. The tornado is supplied with energy. This strong core flow forms and wraps up a superimposed vortex within the mesocyclonic vortex into the tight tornado vortex that will be maintained as long as the jet stream trailing vortex provides the sink and as long as the thunderstorm provides the supporting double vortex structure. The jet stream sweeps the vorticity and core flow downstream to a positive infinity, satisfying the vortex theorems of Lord Kelvin and H. Helmholtz [75]. The generated

133

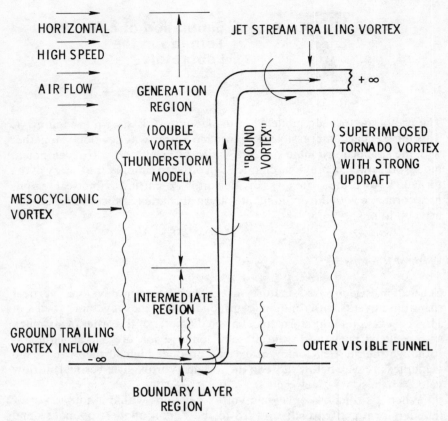

Figure 7-1. Features of a Double Vortex Thunderstorm with a Compressible Flow Tornado Vortex

vortex filament must also proceed to a negative infinity and seeks the
nearest surface, usually the ground.

Intermediate Region. As the tornado vortex extends below the generation
region toward the ground, the core diameter decreases with increasing
pressure (decreasing altitude).

Boundary Layer Region. The tornado vortex, because of viscous action at
the surface, exerts a torque on the ground. The reaction of the torque is to
roll the vortex in an unstable circular motion. The core is bent along the
ground and a ground trailing vortex is formed. This ground trailing vortex
attempts to proceed to negative infinity, satisfying the vortex theorems of
Kelvin and Helmholtz. With the core bent along the ground, the core filling
from the atmospheric reservoir occurs from the direction the core is bent.

The strong core inflow acts like a vacuum sweeper hose. Its track over the ground is in erratic circular patterns if the generation and intermediate regions are stationary. The laboratory model to be described has revealed several features of vortices. When the intermediate region is moving over the surface, a drag is produced on the ground trailing vortex. This drag has a stabilizing influence on the motion of the core. Connecting loops or cycloidal marks are formed if the motion of the intermediate region is slow over the ground. A wavy path or nearly straight path will be produced as the drag increases with increasing translation speed.

Cylindrical Regions

The compressible vortex theory divides the tornado vortex into two non-viscous cylindrical regions: (1) the core region in which the radial component of velocity is negligible; and (2) the free vortex region in which the vertical and radial velocity components are negligible.

In the core region, the following equations apply:

$$p = p_A \left[1 - 0.4 M_{\theta_c}^2 + 0.2 M_{\theta_c}^2 \ (r^2/r_c^2)\right]^{3.5} \tag{7.1}$$

$$p = p_A \left[1 - 0.2(M_w^2 + M_\theta^2)\right]^{3.5} \tag{7.2}$$

$$\mathcal{M} = \int_0^{r_c} 2\pi \rho w r \, dr = \text{constant} \tag{7.3}$$

$$\rho = \rho_A (p/p_A)^{0.715} \tag{7.4}$$

$$v_\theta = v_{\theta_c} r/r_c \tag{7.5}$$

$$M = v/a_A \tag{7.6}$$

In the free vortex region the following equations apply:

$$p = p_A [1 - 0.2 M_\theta^2]^{3.5} \tag{7.7}$$

$$-v_\theta = \Gamma/2\pi r \tag{7.8}$$

In the laboratory modeling of any fluid flow field, the simulation of the fluid dynamic conditions and the boundary conditions imposed on a model must be as similar as possible to the prototype. Many different types of vortices have been created in the laboratory. By varying dynamic and boundary conditions totally different results have been obtained.

The dynamic conditions of greatest importance in modeling a compressible flow are those expressed by Mach number and Reynold's number. In high speed flow the parameter of major interest is Mach number. The Mach number of the two flows should be equal and can be modeled exactly. Reynold's number normally cannot simultaneously be modeled. However, since viscous effects are confined to small boundary layer regions, pressure measurements are almost completely free from Reynold's number effects [79]. Additionally, viscous effects in these thin regions do not change the qualitative test results of other measurements. Frequently the quantitative test data may be satisfactorily corrected for the Reynold's number effect.

The general boundary conditions associated with the prototype tornado that the model must satisfy in so far as possible are:

1. The tornado vortex is generated aloft and drops down from the generation region toward the ground.
2. A horizontal flow of air exists at high altitudes at the top of the thunderstorm to provide outflow and additional energy to the mesocyclonic vortex of the thunderstorm.
3. The tornado vortex of the thunderstorm extends to the ground and travels over the surface at varying speeds. It may skip distances as it travels along the path.
4. The vortex diameter and width of the ground damage vary greatly.
5. At ground level, a decrease in pressure occurs in the vortex region.
6. At some distance from the center of the funnel, the pressure is atmospheric and decreases with altitude.

Laboratory Model

A reasonable simulation of Mach number and boundary conditions of the prototype was provided by the laboratory vortex generator (Figure 7-2). The rotation of an 18-inch diameter wire cage provided circulation at a height well above ground simulation level. The circulation was controlled by varying the speed of the rotating cage. A two-inch diameter hole in the center of the top of the cage provided a variable outflow. The outflow is controlled by varying the pressure at the hole.

This pressure was varied from 1.056 to 0.1 atmospheres. Thus, the rotating cage and decreased pressure simulated variable generation conditions at an arbitrary altitude. The ground surface was simulated by the floor of the laboratory or by a large table top that was adjusted to provide various heights between the generation region and the surface. Also, various angles between the core generation centerline and the surface was established. A moving ground board was used to simulate the tornado translation speed. The square table top area was 2.54 meters on each side. This provided a core to ground radius of approximately one to 750. This is a reasonable simulation of infinity around a vortex. The chief boundary condition that was not simulated to the proper scale was the change in atmospheric pressure with altitude. The flow Mach numbers were simulated.

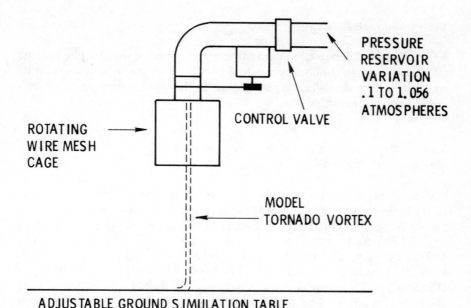

PRESSURE
RESERVOIR
VARIATION
.1 TO 1.056
ATMOSPHERES

CONTROL VALVE

ROTATING
WIRE MESH
CAGE

MODEL
TORNADO VORTEX

ADJUSTABLE GROUND SIMULATION TABLE

Figure 7-2. Laboratory Vortex Generator

Various methods of instrumentation and flow visualization have been used with the model. Ground pressure measurements with a slant water manometer board proved completely unresponsive to the vortex core pressure changes. Standard shock tube pressure measurement equipment provided the rapid response required to accurately follow the core pressure fluxuations as the core moved over the instrument. The Kistler crystal transducers used provided a response time of three microseconds for zero to full-scale reading. A crystal amplifier and oscilloscope with camera were used to record the pressure variation. Tuft guides were used to examine flow directions. The use of water on the surface did not prove to be of much value other than centrifuging out and being very wet. Various colors and types of smoke were used. Their use usually resulted in concluding the day's experiments. Charcoal sparks traced the outer free vortex region. Fire provided some visualization of the core region. The most successful method of visualizing the vertical core region was through the use of dry ice on the table top. The inner core region was clearly visible. The general ground inflow and some of the free vortex were likewise visible.

None of the above methods provided detailed ground flow patterns and track. Oil, glue, paint, china clay, etc. were used for this purpose with little success. Liquid wax proved to be very useful. By varying the viscosity of the wax and changing the wax pattern used on the board, different features of the ground flow patterns and track were visible. A "wax funnel" was also lightly visible. Its diameter was about 50 percent of that of the cage.

Vortex Characteristics

The model has been used to demonstrate a number of specific phenomena associated with the boundary conditions of the tornado vortex. The results were recorded photographically. Some of the phenomena were investigated and the results follow.

*Vortex Produced by Circulation Only (Model
Mesocyclonic Type Vortex)*

The cage rotation alone produced a weak vortex that consisted of a relatively large volume of slowly rotating air. The core diameter was approximately 50 percent of the cage diameter. There was weak updraft in the core region. The free vortex region exhibited an even weaker updraft. The ground pressure changes due to the vortex were negligible. The small suction pressure at the surface is illustrated in Figure 7-3 by the lifting of charcoal particles from the surface. Since the particles have mass, they were thrown outward by the tangential velocity of the vortex. The outer vortex region produced a large visual funnel region with dust or viscous liquid on the surface.

*Introduction of Core Flow into the Circulation Vortex
(Model of Tornado Vortex)*

By controlling the pressure at the hole in the top of the cage, various types of core flow were superimposed upon the circulation vortex produced by the cage. All attempts to produce a vortex by adding a down core flow by a positive pressure at the cage hole were unsuccessful. In fact, the supply of a small pressure over atmospheric into the core at the upper plate disintegrated an existing vortex immediately. This disintegration was so rapid that only three frames of movie camera film were needed to record the phenomena.

A negative pressure at the cage hole triggered a strong upward core flow and a well defined strong localized vortex within the previously weak rotating air mass (Figure 7-4). The well defined upflow core appeared surrounded by a small region of upward spiralling air. The total core region was very small in diameter, approximately 3 percent of the cage diameter. The maximum tangential and vertical velocities were large. There was a marked pressure decrease on the surface immediately under the small central core touchdown region. In the region immediately surrounding the

Figure 7-3. Weak Vortices Produced by Cage Rotation Only. Visualization is by charcoal tracers, top, and by dry ice.

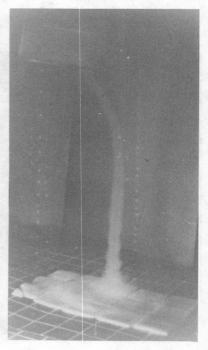

Figure 7-4. Strong Localized Vortex Produced by Cage Rotation and Negative Pressure at the Top of the Cage

core upflow, the flow spirals upward. Outside of the spiraling region, the flow was predominately free vortex flow.

If a below atmospheric pressure was applied at the cage hole without cage rotation, no vortex resulted. However, if the vortex was established by the cage rotation and a below atmospheric pressure, and then the cage rotation stopped, the vortex continued although weaker in strength (Figure 7-5). Thus, when a circulation and outflow were present to initiate the vortex, the strong outflow supplied sufficient energy to drive and sustain the strong core flow and vortex.

The strong vortex was easily blown about in the laboratory by small side drafts. When the core was blown from the center line of the generation region, the core attempted to straighten up as it approached the ground surface. This illustrates the attempt of the vortex to follow the Helmholtz theory, specifying that a vortex in a perfect fluid can end at a solid boundary surface perpendicular to the vortex filament.

Model Tornado Vortex Ground Pressure Measurements

Ground pressure measurements in the immediate vicinity of the model tornado vortex core have been markedly below atmospheric pressure as contrasted with the negligible pressure changes measured with cage rotation only. Figure 7-6 shows typical pressure traces of the vortex core when there was no relative movement between the generation region and the ground board. Using the fast response instrumentation, the pressure measurements showed a highly localized low pressure region. Significant features of the pressure measurement data follow.

1. No positive pressure measurements were observed in the vicinity of the vortex.

2. Without relative motion between the cage and ground board, the core center passed directly over the transducer less than once each 10 seconds of exposure time. The transducer, 1.0 m in diameter, was located in the center of the damage area (Figure 7.10). A minimum core pressure of 0.577 atmospheres was measured.

Using Equations (7.1) and (7.2), this pressure gives a maximum tangential Mach number, M_{θ_c}, of 0.6029 and a maximum core flow Mach number of 0.8526. The minimum core density is $0.6749\,\rho_A$. The theoretical pressure, density, and Mach number distributions are plotted in Figure 7-7. Pressure data points from Figure 7-6 are shown also in Figure 7-7.

Using an effective core Mach number of 0.65 and p/p_A of 0.65, the dynamic pressure may be computed as

Figure 7-5. Strong Vortex Maintained by Negative Cage Hole Pressure and No Cage Rotation

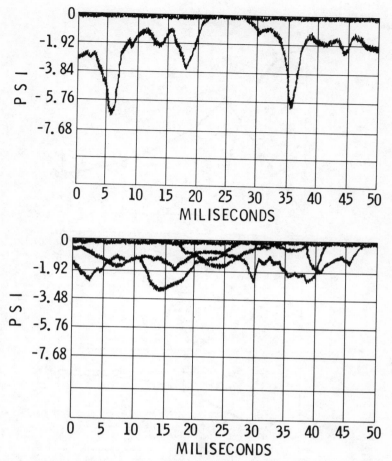

Figure 7-6. Pressure Trace at a Fixed Point on the Surface (Ten-seconds Exposure)

$$q = \frac{\gamma p M^2}{2} = \left(\frac{1.4}{2} \right) (0.65 p_A)(0.65)^2 = 0.1922 p_A$$

T. R. Turner [82] shows the lift coefficient for a 1965 Ford Galaxie two door hardtop as varying from 0.4 to 0.9, depending on the angle of yaw. The drag force coefficient was 0.5. Using the value of 0.4 for the lift coefficient and the above dynamic pressure, the lift and drag forces are

$$L = q C_L A = 0.192 p_A \frac{(0.4)(3.466)}{\sqrt{1 - 0.65^2}} \frac{(64)}{9} = 5{,}270 \text{ pounds}$$

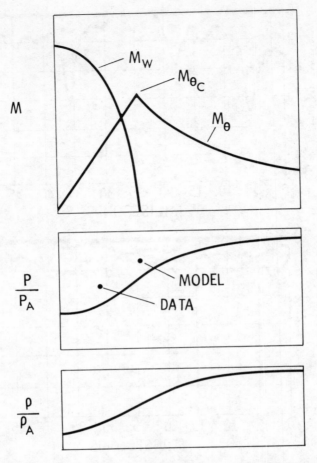

Figure 7-7. Comparison of Computed and Measured Pressures—Tornado Vortex Simulation

$$D = qC_DA = 0.192p_A \frac{(0.5)(3.466)}{\sqrt{1 - 0.65^2}} \frac{(64)}{9} = 6,588 \text{ pounds}$$

This would be more than sufficient lift to make the automobile "fly." An effective core flow Mach number of this order of magnitude not only would explain the flight of many unusual objects but also the roaring sound associated with the tornado funnel.

T. T. Fujita [22] reported the movement of a chain of 15 freight cars on a railroad siding at Lubbock, Texas. The final disposition of the cars indicate the path of a strong "suction spot." Two cars were carried as far as 70 meters off the tracks. J. T. Matthews and W. F. Barnett [83] show the coefficients for an empty automobile rack car in a group of cars as $C_L = 1.686$; $C_D = 2.08$; and $C_S = 1.81$. These values are for a yaw angle of 30°. Using the model effective core dynamic pressure, this would give

$$L = \frac{0.192 \, p_A \, (1.686)(125)}{\sqrt{1 - 0.65^2}} = 112{,}670 \text{ pounds}$$

$$D = \frac{0.192 \, p_A \, (2.08)(125)}{\sqrt{1 - 0.65^2}} = 138{,}999 \text{ pounds}$$

$$S = \frac{0.192 \, p_A \, (1.81)(125)}{\sqrt{1 - 0.65^2}} = 120{,}956 \text{ pounds}$$

Although the car configurations were not reported, such forces would appear to provide a reasonable explanation of the occurrence. The lifting of tops of grain storage bins has been reported [22]. The tops were of concrete 10 inches thick. Wind tunnel tests have indicated that negative pressure coefficients of 1.4 are not unusual for the top of a flat sharp-edged roof. If the bins were not completely vented simultaneously with the tornado core striking them, this would give a lifting force per square foot of the tops of

$$L = \frac{0.192 \, p_A \, (1.4)}{\sqrt{1 - 0.65^2}} = 748.5 \text{ pounds}$$

Ten-inch thick concrete weighs about 120 pounds per square foot. Although venting would occur as the top began failing, the raised top would catch the wind and be peeled back. The "flying" of a 16-ton empty fertilizer tank has been reported [22]. Since the dimensions of the tank are not known, an estimate of the forces on it cannot be calculated.

3. With relative motion between the table and the ground board sufficient to produce a nearly straight wax ground track, the pressure change was extremely localized. No significant pressure changes were detected when the distance from the center of the transducer to the centerline of the track was greater than one track width. Limited testing has shown that the maximum negative pressures measured with the moving board were about 0.4 of those with the stationary board. This indicated that the center of the core of the ground trailing vortex was approximately one-fourth of the core diameter above the surface.

Ground Tracks

By using a film of wax on a ground board, the ground trace of the core region and the ground flow pattern were recorded (Figures 7-8 through 7-15).

When there is no relative motion between the cage and the ground

Figure 7-8. Ground Trailing Vortex Core Track—No Relative Motion Between Cage and Ground (One-half Second Exposure)

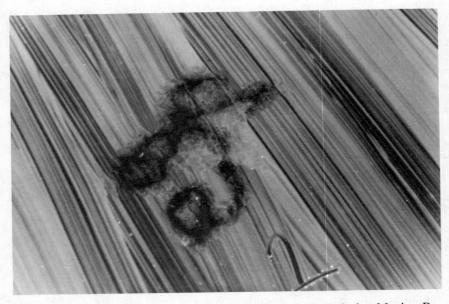

Figure 7-9. Ground Trailing Vortex Core Track—No Relative Motion Between Cage and Ground (One-second Exposure)

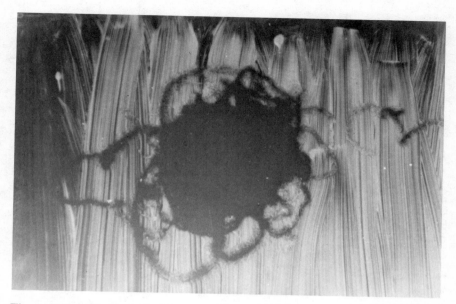

Figure 7-10. Ground Trailing Vortex Core Track—No Relative Motion Between Cage and Ground (Twelve-seconds Exposure)

Figure 7-11. Ground Trailing Vortex Core Track—Slow Relative Motion Between Cage and Ground

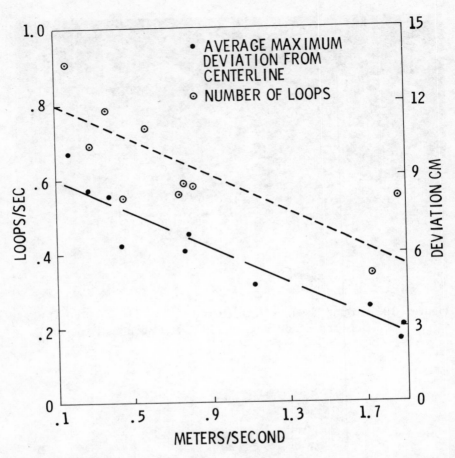

Figure 7-12. Loops and Track Deviation as a Function of Translation Speed

surface (Figures 7-8, 7-9, and 7-10), the ground trace of the core was unstable and prescribed random paths over the surface. The circulation around the updraft core provided a torque on the core as it contacted the surface. This torque caused the core to bend and lie along the surface. The core section along the surface filled from one general direction. From the figures it can be seen that the high velocity air entering the core left a distinct ground track and flow pattern. As the time of surface exposure to the tornado model vortex increased, the central damage region expanded as the wandering core recrossing its path a number of times. Also evident were the random excursions beyond the central damage region.

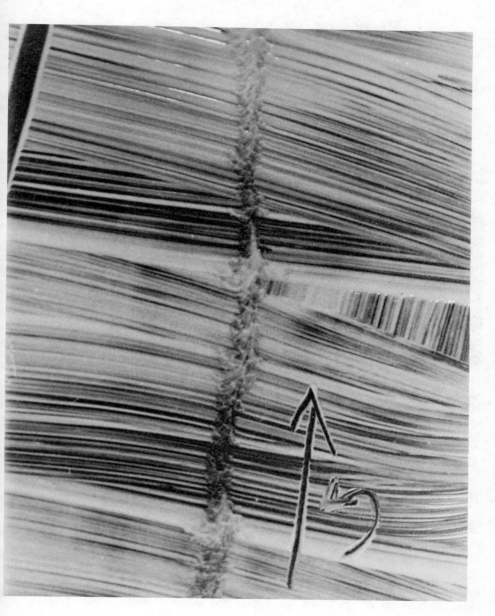

Figure 7-13. Surface Wind Effects from Laboratory Tornado Model with Rapid Translation Speed

Figure 7-14. Ground Trailing Vortex Core Track—Skip Distances

Figure 7-15. Ground Trailing Vortex Core Track and Flow Pattern

With slow relative motion between the core and the ground surface (Figure 7-11) the core described various loops and arcs along the surface. These shapes were a function of the translation speed, vortex strength in the generation region, and the distance from the generation region to the surface. The ground leg portion of the vortex was partially stabilized by the drag on the core from the movement of the core over the ground (Figure 7-12). The inflow to the core at the ground was in the instantaneous direction of the ground leg portion of the vortex. The core moved in a cyclonic motion in describing the loops with cyclonic cage rotation. With rapid movement of the core relative to the ground surface, the drag stabilized the ground trace (Figure 7-13).

The core trace width was a function of the distance between the surface and the generation region, the cage rotation speed, and the outflow velocity. Keeping the latter two constant, the core trace width decreased as the distance between the surface and generation region increased. As this distance was increased, the trace began to skip along the track (Figure 7-14).

The ground trailing vortex core flow pattern on the ground was particularly evident in Figures 7-8, 7-13, and 7-15. The central areas showed the effect of the high velocity core inflow. This ground flow moved very rapidly in the direction of core movement. The maximum flow speed occurred about 2/5 of the track width from the right side of the track with cyclonic rotation. The width of the maximum flow area is about 1/5 the width of the damage track. In the damage track the wax movement in the area to the right of the maximum flow area was inward and in the direction of translation. On the left side of the maximum flow area, the wax movement is inward and in the direction of translation also. The area is wider than the corresponding area to the right of the maximum flow area. The flow is less uniformly severe and more circular. The flow lines correspond quite clearly with inflow streaks that have been called suction spots [22] and were assumed to be due to smaller vortices within the main vortex. However, they occur in the laboratory model because of the inflow patterns at the base of the vortex. The relatively small effect of the tangential winds of the vortex can be seen at the edge of the ground trailing vortex core flow pattern. In examining the movement of wax, the maximum movement in the core center region was from two to four times that in the rotational region. Thus, the predominant or most serious flow from a damage point of view was the ground flow into the core in the direction of the core movement.

Correlation with Ground Observations

The wax paths of the tornado model vortex present excellent correlation with the ground damage patterns observed by the authors. In both the severe damage region indicated that the damage was caused by a high velocity wind blowing in the instantaneous direction of the core movement. The path of the Lubbock tornado (Figure 3.6), developed from wind damage observations by the authors, corresponded to typical model tracks. Most of the excellent detail of the wind streaks produced by the Lubbock tornado correspond with the results from the laboratory model. Therefore, the observed tornado ground tracks and wind streaks may be explained by these ground trailing vortex and inflow patterns.

Although the magnitude of the pressure change of the model is much larger than recorded for tornado passage, it should be noted that the actual pressure has never been recorded with a high response rate instrument. The significant pressure changes on the model occurred within the wax damage path. The model pressure pattern correlates well with the Topeka tornado pressure pattern. The Weather Bureau of Topeka was located about 250 feet to the left of the Tornado tornado track (June 8, 1966). The pressure trace showed a decrease in pressure four to eight minutes before the arrival of the tornado, a rising pressure just before passage, a very rapid dip at passage, a rapid rise, followed by about five minutes of decreased pressure. Considering the Topeka Weather Bureau location, it should have been subjected to, first, the mesocyclonic low pressure, then higher pressure in the low velocity region just before the high velocity core flow passage. The latter would cause a very fast dip in pressure. Following such a rapid dip, the instrument probably would overshoot and then settle down on the mesocyclonic low pressure. Qualitatively, the results are similar to that of the model at the edge of the damage path.

T. T. Fujita [22] suggests that the major effects of the tornado are inside suction spots with horizontal dimensions one order of magnitude smaller than the tornado core. The rate of change of pressure is also suggested to be an order of magnitude greater than previously computed from pressure profiles. It is noteworthy that the diameter of the compressible vortex core flow in the laboratory was approximately one-sixteenth of the diameter of the visible "wax funnel." The core pressure changes measured in the laboratory were approximately $0.3p_A$ within the distance of the core radius. Figures 7-6 and 7-16 show the existence of large pressure gradients. Estimates of pressure differentials in tornadoes [74] vary from 200 psf to 2,000 psf. The model measured differential was 895 psf.

Conclusions

The laboratory vortex generator provided excellent modeling of the prototype tornado vortex. Essential boundary and Mach number conditions were modeled. It demonstrated that a sink aloft was required to establish a strong upward core flow to trigger and maintain a strong tornado type vortex within the weaker mesocyclonic vortex.

The strong core flow tracks in the wax showed the ground inflow pattern to the core as a very strong flow in the general direction of the instantaneous direction of the ground trailing vortex, as in the theoretical model. The path of the core in the wax was also as presented in the theory. The laboratory tornado vortex paths and inflow patterns in wax provided an excellent explanation of observed tornado damage patterns.

The model ground pressures indicated a marked decrease in pressures in the core region. Visual observations indicated high flow velocities in the core region. Both show the region to be highly localized. Although comparable field measurements have never been made in the tornado core region, it is considered that these conditions do exist in the tornado vortex core. A scientific investigation of the tornado vortex is needed to further validate the tornado model and quantitatively determine the flow field.

The model demonstrated that the strong tornado type vortex was completely dependent upon a strong sink aloft for its existence. Cutting off the sink destroyed the vortex. Supplying pressure to the weak circulation vortex destroyed it. This suggests that the tornado may be destroyed by supplying a pressure to the core in the generation region.

8 Synoptic Conditions for Tornado Development

Atmospheric Conditions Leading to Tornado Development

A number of specific atmospheric conditions are necessary for the development of supercell thunderstorms that are responsible for most tornadoes. A small number of tornadoes occur in thunderstorms developed in the leading edge of hurricanes. Others sometimes occur in the southern United States from smaller thunderstorms where the main generation mechanism is very warm moist unstable air. Most tornadoes develop from large thunderstorms, however, as a result of several combined atmospheric conditions.

Jet Stream and Upper Air Divergence

Tornadoes are associated with the jet stream and with areas of horizontal divergence at 500 mb and above. When a region of divergence exists in the upper troposphere this means there is net outflow in the horizontal plane. This contributes to updrafts below the area of divergence and allows developing thunderstorms to grow large with higher velocity updrafts. In addition, the speed of the jet stream imparts energy to the thunderstorm. If the velocity of the upper air exceeds the speed of the thunderstorm, which is typically about 30 mph, it provides a force to move the thunderstorm along as well as adding circulation to it because of flow around the thunderstorm which blocks the main stream of air.

Moisture Tongue and Low Level Jet

Tornado activity is usually associated with warm humid surface air flowing northward from the Gulf of Mexico. The dew point temperature is normally greater than 55° F and frequently a narrow band of warm humid air flows northward from the Gulf of Mexico on days with tornadoes. This occurs because of the surface pressure patterns with anticyclonic flow around surface high pressure and cyclonic flow around lows. Squall lines frequently develop along the western boundary of the moisture tongue. This boundary separates the warm humid Gulf air from the hot dry air that

155

normally comes from the southwestern desert region. This boundary (dry line) represents an instability line since the hot humid air is less dense than the hot dry air that is normally moving northeastward. A low level jet of higher velocity air than normal also frequently accompanies the moisture tongue and contributes to the proper wind shear environment for tornado development.

Upper Air Inversion

An inversion layer, where the temperature increases with height, is a very stable layer since the colder more dense air is on the bottom, decreasing the tendency for vertical motion. On days with tornado activity, an inversion normally exists because of the warm humid air flowing northward from the Gulf and the cool dry air flowing eastward over the Rockies. An inversion exists at the boundary between these two air masses at about 850 mb. The effect of the inversion layer is to dampen out small cumulus buildup that originates within the warm humid air below 850 mb until a cloud is buoyant enough to penetrate the stable inversion layer. When this happens the cumulus cloud is able to grow much faster than if the inversion layer had not held down previous development.

Unstable Atmosphere

The atmosphere is very unstable when tornado outbreaks occur. There is an increased tendency for vertical motions to occur because of warm air near the earth's surface. A number can be placed on the stability by several techniques. One of these is the lifted index [84]. The lifted index is determined by comparing the measured temperature at 500 mb with the temperature of a parcel of air having the forecasted maximum temperature for the day and the average humidity within the lowest 100 mb layer. Air with this temperature and humidity is theoretically lifted, dry adiabatically, until saturated, then moist adiabatically to 500 mb for comparison. If the temperature of the lifted parcel of air is warmer than the environment at 500 mb a negative value results and indicates an unstable atmosphere. The magnitude of the negative number is related to the degree of instability of the atmosphere.

An unstable atmosphere arises because of surface heating from solar radiation and from contrasting air masses when cool dry air overrides warm moist air. Most of the time both of these work together so that tornado development reaches a peak in the late afternoon hours. Sometimes the contrasting air masses are so different that surface heating is not required to trigger thunderstorm development, resulting in tornadoes at other hours of the day or night. Occasionally the air is so humid and warm, especially in the southern part of the United States, that tornadoes develop without contrasting air masses.

Proper Wind Profile

It has been observed that large differences in wind direction and velocity exist during tornado activity. There has been little agreement, however, on the role of such differences in support of severe thunderstorms. On the basis of the double vortex thunderstorm model the required wind profile is one that can provide upper level winds that are opposite and of similar magnitude to low level winds relative to a moving thunderstorm. The forward speed of a thunderstorm normally subtracts from the measured upper level winds and adds to the low level winds because of the opposite directions of flow. Since a thunderstorm can move at various speeds depending on its circulation and the environmental winds, the proper wind profile is one that allows a balance between the relative winds at low and mid-levels for a particular storm speed and direction.

Midlatitude Cyclone

Tornado development is most common in the warm air sector of a mid-latitude cyclone. The most common region of squall line development is within the warm air mass along the dry line that separates the warm humid Gulf air (*mT*) from the hot dry desert air (*cT*) coming from the southwest. Sometimes the hot dry desert air is absent and then squall line development occurs predominately along and in advance of the cold front that separates the cold air mass (*cP*) from the Gulf air. Other locations where tornadoes sometimes occur are deep within the warm air mass, along the warm front, and near the center of the mid-latitude cyclone.

Squall Lines and Isolated Thunderstorms

Tornadoes commonly occur from thunderstorms that are part of a squall line composed of several thunderstorm cells. Within a squall line tornadoes are more likely from the largest thunderstorms within the line. These are commonly the southernmost cell in a line, the cell near a bend in a bow-shaped squall line, or a cell that intersects the cold or warm front.

Tornadoes also frequently occur from large, isolated thunderstorms. These normally develop within the warm air mass.

Location of Tornado Activity

When the seven atmospheric conditions just described coincide over a given geographical location, tornado activity is likely. As indicated in Figure 8-1 this occurs in a fairly well defined area so that tornado forecasts can be issued based upon the combination of atmospheric conditions that is likely to lead to tornado development. In Figure 8-1 the divergence of streamlines at the 500 mb level is shown by the band surrounding the jet stream that is flowing from west to east with cyclonic curvature in this example. Divergence at 500 mb is indicated by the widening of streamlines over Arkansas. The jet stream is above the 500 mb surface and normally is located at about 300 mb. At the surface the cold front in this example would be pushed along rapidly by the upper air flow. The edge of the moisture tongue is shown by the dashed line and contains warm humid air drawn up from the Gulf. A wedge of continental tropical air from the Desert Southwest separates the maritime tropical air from the continental polar air behind the front. With an inversion and negative lifted index tornado activity would be most likely over Arkansas as the dry line (western boundary of the moisture tongue) triggers squall line development.

Synoptic Patterns for Four Major Tornado Days

Four severe weather occurrences were selected for comparison. Two of the most damaging single tornadoes in history were the Topeka tornado on June 8, 1966, and the Lubbock tornado on May 11, 1970. Both resulted in damages amounting to over $100 million. The two most severe tornado days since the tristate tornado in 1925 were April 11, 1965, and April 3, 1974. Major tornado damage and many injuries and deaths occurred in several states on both these days. In the 1965 outbreak 258 deaths and 3,148 injuries occurred in six different states. Property damages amounted to about $238 million from at least 37 separate tornadoes. These statistics were surpassed on April 3, 1974 when more than 100 tornadoes occurred in 11 states. These tornadoes left more than 4,000 people injured, 329 deaths, and about $700 million damage to property.

A comparison of the synoptic patterns for the major tornado days in 1965 and 1974 can be made from Figures 8-2 and 8-3. Extreme divergence at 500 mb is indicated on April 3, 1974 (Figure 8-2), for the United States south of the Great Lakes. This was combined with a strong mid-latitude cyclone, a distinct dry line with very warm humid unstable air east of the dry line. Intense tornado activity occurred in Ohio, Indiana, Kentucky, Tennessee, Alabama, Georgia, North Carolina, Virginia, West Virginia, Michigan, and

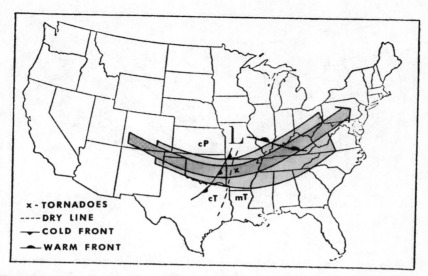

Figure 8-1. Combination of Several Atmospheric Factors That are Important in Determining Where Tornado Activity Will Occur. The west to east arrow shows the jet stream location, with the wider band indicating 500 mb divergence.

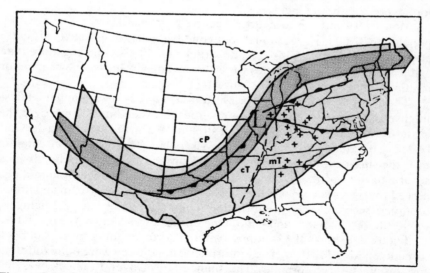

Figure 8-2. Synoptic Conditions on April 3, 1974, at 7:00 P.M. CDT, Showing Location of the Surface Features and Upper Air Patterns Responsible for the Most Tornado Activity in Modern Times

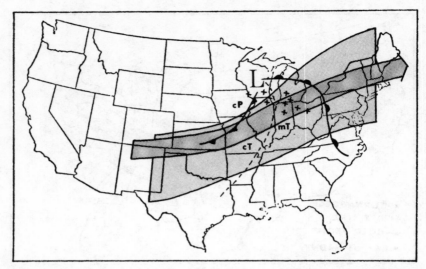

Figure 8-3. Synoptic Patterns on April 11, 1965, at 7:00 P.M. CDT, Showing Atmospheric Features Corresponding to Major Tornado Activity

Illinois beneath the area within the atmosphere where all the components for severe weather coincided.

Many very intense tornadoes occurred on this day. In Xenia, Ohio, population 27,000, a large funnel about one-half mile wide destroyed nearly 50 percent of the city. This tornado struck at 4:40 P.M., injuring more than 1,100 people. Many other cities were also invaded by massive tornadoes or multiple funnels as shown in Figure 1-13.

The synoptic situation on April 11, 1965 (Figure 8-3) was similar but less intense than in 1974. Large-scale divergence existed at 500 mb over the eastern United States beneath the jet stream. Warm, moist air from the Gulf of Mexico extended as far north as Wisconsin. The cyclonic circulation near the surface low and convergence of surface air at the dry line, which separated the warm dry wedge of air to the west of the warm humid air, served to trigger severe weather activity beneath the jet stream and upper air divergence area. Severe tornadoes occurred in Wisconsin, Michigan, Iowa, Illinois, Indiana, and Ohio. One of these was shown as Figure 1-9.

Figure 8-4 shows the synoptic patterns corresponding to the Lubbock tornado on May 11, 1970. The atmospheric conditions were quite different for this very large tornado. The thunderstorm over Lubbock was not initiated by the usual large-scale atmospheric conditions. Very little upper air divergence is indicated and the jet stream is displaced north of Lubbock. The tornado in Lubbock occurred from an isolated thunderstorm within the

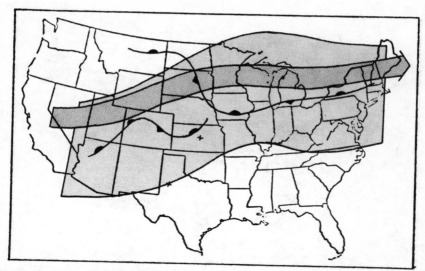

Figure 8-4. Synoptic Conditions on May 11, 1970, at 7:00 P.M. CDT, Corresponding to the Lubbock Tornado

warm air mass. This type of thunderstorm also affects the other Gulf states. A tornado also occurred near Ellsworth, Kansas, ahead of a stationary front on May 11, 1970. The Lubbock tornado occurred just after dark and was not photographed, but the Ellsworth tornado development was photographed as shown in Figure 1-3.

The Topeka tornado occurred with more typical atmospheric conditions for tornado development within the Great Plains. Figure 8-5 shows that upper air divergence, the jet stream, and a surface low pressure system all collided over eastern Kansas at 7:15 P.M. to produce a massive funnel (Figure 2-1) in Topeka similar to the Lubbock tornado. The upper air support (jet stream and upper air divergence) were greatest over Kansas, although some severe thunderstorms occurred in Oklahoma just ahead of the dry line. The Topeka tornado occurred near the center of the mid-latitude cyclone where it could take advantage of the rotation and upper air support. The lifted index was calculated as a measure of atmospheric stability (Figure 8-6) on June 8, 1966. At 6:00 P.M. the atmosphere was most unstable over Topeka (-6) with a band of unstable air extending southward through Oklahoma. Additional unstable air occurred in the southeastern United States, but did not have sufficient upper air dynamics for support of tornado activity.

It is apparent from comparison of the atmospheric conditions for these four major tornado outbreaks that in three of the four cases the atmospheric

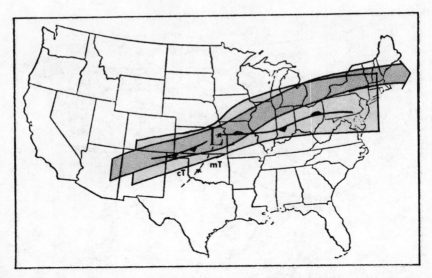

Figure 8-5. Synoptic Conditions on June 8, 1966, at 7:00 P.M. CDT, Corresponding to the Topeka Tornado

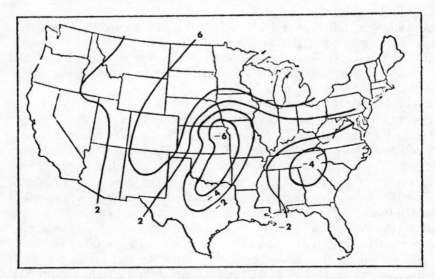

Figure 8-6. Atmospheric Stability on June 8, 1966, at 7:00 P.M., CDT, as Calculated by the Lifted Index

patterns were quite similar, emphasizing the importance of the combined effects of upper air support over warm humid surface air that is being replaced by more dense warm dry air. The wind profile under this condition consists of south surface winds with more westerly upper air currents.

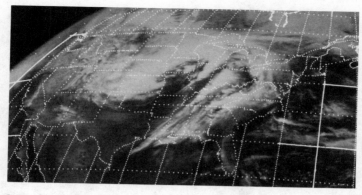

Figure 8-7. Cloud Patterns Over the United States at 2:55 P.M., CDT, on April 3, 1974, as Viewed from ATS-3

Thermodynamics are important, since most of the severe weather activity occurs in the warm moist air, and may be sufficient to compensate for a lack of strong upper air support as occurred in the Lubbock storm in 1970.

Satellite and Radar Observations

Satellite photographs obtained from ATS-3 on April 3, 1974, show the squall line activity. Figure 8-7 taken at 2:55 P.M. shows lines of convective activity extending through most of the states affected by severe weather. The mid-latitude cyclone was centered over Missouri at this time. Cloud patterns over Kansas, Oklahoma, and New Mexico show the movement of the incoming continental polar air mass. Southwesterly flow within the warm air mass is evident in the eastern United States. Three major squall lines contained thunderstorms that were supporting tornadoes. One of the lines extends southwestward across Illinois and cannot propogate any further southwestward into Missouri because of insufficient humidity in the air. This squall line contained large thunderstorms that produced tornadoes across Illinois and northern Indiana. The second squall line covers much of Indiana at the time of the photograph and stretches southwestward into Kentucky and Tennessee. This line produced severe tornadoes in these states as well as in Ohio and Alabama as it moved eastward. The third major thunderstorm area is to the east covering eastern Tennessee and Kentucky. It produced tornadoes in northern Georgia and the western part of North Carolina as well as in Tennessee.

The single thunderstorms that produced tornadoes in Lubbock, Texas, and Ellsworth, Kansas, can be seen in the ATS-3 photograph taken on May 11, 1970 (Figure 8-8). It would be very useful if such thunderstorms had a unique signature from space, but the tornado is so small and the thunder-

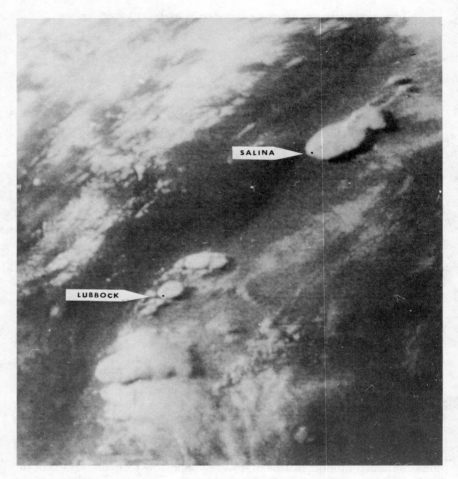

Figure 8-8. Thunderstorms Over Texas and Kansas That Produced the Lubbock, Texas, Tornado and the Ellsworth, Kansas Tornado on May 11, 1970 as Viewed from ATS-3 at 5:30 P.M., CDT (Photograph courtesy of National Environmental Satellite Services, National Oceanographic and Atmospheric Administration)

storm so similar to nonsevere thunderstorms that more detailed information is needed. One of the most promising signatures from space is the explosive development of tornado producing thunderstorms revealed by time series satellite photographs.

Radar offers another method of remote sensing of tornado activity. Unique observations of several severe thunderstorms in action at once

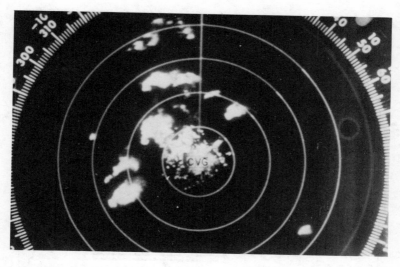

Figure 8-9. Radar Echoes Showing Severe Thunderstorms in the Cincinnati Area on April 3, 1974, at 3:20 P.M., CDT

were obtained by the Cincinnati National Weather Service on April 3, 1974 (Figure 8-9). These four major thunderstorms traveled toward the northeast as shown in Figure 8-10. Storm D, which struck Xenia, Ohio, was downwind from storm C, which moved through Cincinnati. Storms A, B, and C had distinct hook echoes at 15:20 CDT while storm D was still developing at this time and formed a hook echo several minutes later. The forward speed of these storms varied somewhat with stage of development, but was about 50 mph through their severe stages.

Storm A produced three tornadoes, which left damage paths 12, 18, and 21 miles in length through Shelby, Hancock, and Randolph counties (Figure 1-13). According to Storm Data [24] the second of the three tornadoes developed by this thunderstorm was 1,000 yards in width and caused a woman to go into premature labor that resulted in a stillborn baby.

Storm B also developed three different tornadoes, which left damage paths 26, 29, and 32 miles in length. These funnels were also slightly larger than average since they were 440 yards wide. This storm killed three people—two located in mobile homes.

Storm C was likewise a repeating tornado producer. The first funnel traveled 67 miles and was 700 yards across. The second was 1 mile wide and traveled 30 miles. The third funnel was 300 yards wide and traveled 32 miles. Multiple funnels were reported for part of this distance. This thunderstorm was responsible for 17 deaths and 370 injuries in southern Indiana. Storm C continued into Ohio where twin funnels crossed the river

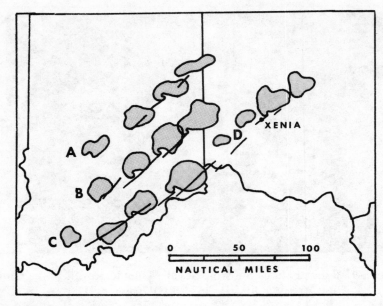

Figure 8-10. Radar Echoes and Tornado Damage Paths for the Four Severe Thunderstorms That Produced Tornadoes in South Central Indiana (cells A and B), in Southern Indiana, Across Cincinnati, Ohio (cell C), and in Xenia, Ohio (cell D)

from Kentucky into Indiana, then back into Kentucky and finally into Cincinnati. Two additional funnels damaged Cincinnati suburbs.

Storm D developed northeast of storm C and almost directly in line with its direction of travel. This cell developed a large severe funnel, which traveled 16 miles through Greene County with a damage path width of 1,300 yards. It destroyed about one-half of Xenia, killing 34 people and injuring 1,150 others. Most of the central business district was destroyed as well as two nearby colleges. Heavy rain and golf ball sized hail accompanied the tornado.

The wind profile measured at Dayton, Ohio, at 6:15 P.M. was used to calculate the winds relative to the moving thunderstorms on April 3. Thunderstorms traveling at a speed of 60 percent of the mean wind would have been moving at 51 mph, which is close to their observed speed. This would have resulted in relative winds as shown in Figure 8-11 for echoes from storms B (upper right) and storm D (lower left). The relative winds from 550 to 300 mb were 51 mph at 117°, which opposed the low level relative winds at 37 mph and 249° for storms moving 10° to the right of the mean winds. The opposing components of the winds would have been very close in magnitude and properly directed for development of the double

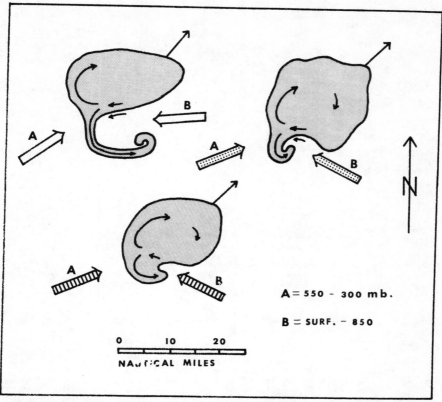

Figure 8-11. Orientation of the Observed Hook Echoes in Comparison to the Relative Winds Calculated from Rawinsonde Measurements for Three Storms: Topeka, 1966, in upper left; Xenia, 1974, in lower left; and cell B (Figure 8-10), 1974

vortex structure as indicated in Figure 8-11. This has been true of all other cases that have been investigated. The hook echo that was observed on the Topeka radar at 7:03 P.M. on June 8, 1966, is also shown in Figure 8-11 (upper left). This echo had a protruding hook that emphasizes the importance of opposing relative winds for strong tornado development. These relative winds were calculated from the Topeka radwinsonde data for a storm moving 10° to the right of the mean wind and at 50 percent of the speed of the mean wind.

9

Energy-Shear Index for Forecasting Tornadoes

The forecasting of tornadoes has been of considerable interest for some time. E.J. Fawbush and R.C. Miller [85, 86] analyzed and classified tornado proximity soundings into three air mass types, noting the general characteristics of temperature, wind and frontal conditions for each type. Several indices [84, 87, 88] have been developed that are concerned only with the thermodynamic conditions of the atmosphere necessary for tornado development. Each relates temperature and moisture characteristics to the potential instability of the atmosphere. Darkow's Energy Index (EI) was developed from the basic equation for the total energy of a unit mass of air. The Showalter Index and Lifted Index consider only the contribution of ascending, warm, low level air to the development of a severe thunderstorm; however, Darkow's Energy Index also takes into account the descending, potentially cold air, the importance of which has been emphasized by several investigators [26, 32, 89]. The EI is calculated by combining the change in temperature, change in mixing ratio, and depth of the layer between the 500 and 850 mb levels.

Necessary environmental wind conditions for tornado development has been considered by only a few people. It has been observed [86] that winds veer with height in tornado proximity soundings. It has also been concluded [90] that the most favorable area for severe thunderstorm development was "in the area of maximum anticyclonic shear in the shear wind field." It has also been hypothesized [91] that the presence of a large vertical wind shear between the low and middle levels was important in tornado formation. This hypothesis has been used [92] to formulate a Tornado Likelihood Index (TLI), combining the factors of vertical wind shear and a thermodynamic parameter, cumulus potential buoyancy. As a measure of cumulus potential buoyancy, the maximum change in psuedo-equivalent potential temperature between the 500, 700, and 850 mb levels and the surface was used.

It was suggested in previous chapters that the mechanism for tornado genesis was the development of a double vortex thunderstorm as a result of 180° veering of the relative winds between the surface and middle levels. The double vortex thunderstorm model was developed on the basis of potential flow theory combined with observed characteristics of severe thunderstorms. An essential feature of the model involves the incorporation of low level moist air into the front of the storm where it rises and collides with the mid-level flow from the opposite direction. The model suggests a closer examination of the relative wind shear profile. On the basis of this model it should be possible to develop a wind parameter corresponding to tornado formation. Because no tornado forecast index has been formulated that combines such a relative wind parameter with thermodynamic considerations, it is the purpose of this chapter to investigate this possibility.

Data Acquisition

The data used in this study were obtained in three parts. First, an attempt to get tornado proximity soundings was made using *Storm Data* [24], concentrating primarily on the years 1965-70. A tornado proximity sounding was defined as a sounding that was within the warm air sector and less than 120 miles of a confirmed tornado touchdown. The time requirements were that the sounding occurred within two hours before the tornado or no more than one-half hour after the first report of a tornado. Up to a half-hour after the tornado was allowed because the reported times of tornado occurrences are usually only approximate and the conditions of the atmosphere that spawned the tornado are probably still operating. Twenty-five soundings were obtained in this manner; it was difficult to find soundings that fell into this category as upper air stations are usually too far away in both time and distance from reported tornado touchdowns.

Second, an attempt was made to obtain proximity soundings and soundings for the same time but located over two hundred miles away from a tornado occurrence. The third acquisition of data was for the April 11, 1965, tornadoes, the Topeka tornado on June 8, 1966, and a large tornado in Minnesota on June 13, 1968. Data from several stations were obtained so that conditions over a wide area on a day of a major tornado could be examined. In addition to proximity soundings for each, 12 additional soundings were obtained for the Palm Sunday tornadoes, 6 for the Topeka tornado, and 5 for the Minnesota tornado. Fifty-nine soundings were, therefore, used in this study, 27 proximity soundings, 12 precedence soundings, which were soundings taken in the warm air ahead of the cold front but removed from it in either time or distance, and 20 nonproximity soundings.

Development of the Shear Index

Basis of the Index

The winds relative to a moving thunderstorm are not the same as the winds

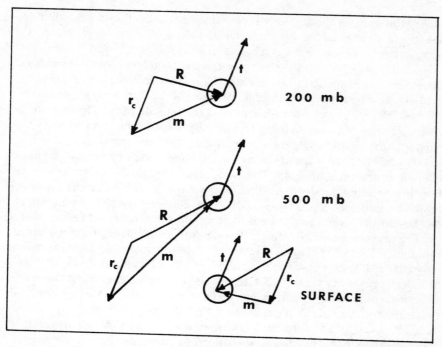

Figure 9-1. Vectors for a Particular Tornado Proximity Sounding Showing the Wind (R) Relative to a Moving Thunderstorm as Determined by the Movement of the Storm (t), Which Creates a Wind (r_c) Opposite to the Direction of Movement of the Storm. The combination of (r_c) with the measured winds (m) relative to a fixed point results in the relative winds (R) for a moving thunderstorm

measured by a radiosonde (Figure 9-1). The measured winds (m) usually consist of winds with a southerly and an easterly component in the layer below 850 mb that veer rapidly to acquire a westerly component at mid-tropospheric levels. The wind relative to the cloud if the air were calm is shown in the figure as (r_c); it is the result of the motion of the cloud itself and is equal in magnitude and opposite in direction to the cloud movement (t). The developing cloud will move at approximately the same direction as the environmental wind, which is on the average out of the southwest. The vector addition of m and r_c therefore, yields a low level relative wind (R) with a strong easterly component and a mid-level relative wind with a westerly component. Therefore, the relative winds between these two levels oppose each other.

In the double vortex thunderstorm model two vortices develop, one cyclonic and the other anticyclonic. If the relative winds oppose each other, the moisture laden, low level relative winds entering the thunder-storm are carried along by the updraft to the 500 mb level. Here, the opposing relative winds act as a barrier; as the low level wind collides with this barrier, it is split into various parts, some traveling around the sides of the storm and some traveling upward. Thus, two vortices are created, one cyclonic and the other anticyclonic, because of the flow around the updraft barrier. The mid-level air travels around the outside of the storm with the rising, low level winds helping to intensify the vortices. As long as this barrier to the incoming winds is maintained, the thunderstorm can continue to build and intensify; the stronger the cyclonic vortex becomes, the more likely a tornado will develop. The basic requirement, then, that allows the flow inside a thunderstorm to simulate a solid cylinder is that the low level relative winds are directly opposed by the mid-level relative winds. This is the key to the wind requirement for tornado formation and forecasting.

The relative winds affecting a thunderstorm depend on the storm direction and velocity. Frictional effects and rotation cause the cloud to move only at 50 to 80 percent of the mean environmental winds. Veering of a split thunderstorm by as much as 30° to the right or 50° to the left of the mean environmental wind direction has been reported [34]. It was observed that tornadoes can occur in either the left or right moving storms. It has also been observed [66] that storms late in their life veer up to 60° to the right. In choosing the storm speeds and directions to calculate the relative winds, therefore, it was ncessary to try a wide variety of values.

Wind data used in this study is currently reported for the surface and the following significant pressure levels: 850, 700, 500, 400, 300, 250, 200, 150, and 100 mb. The mean environmental wind velocities were calculated, therefore, as other investigators [66] have done, by finding the vector mean of the 850, 700, 500, and 300 mb reported winds. Using this mean wind velocity, the relative wind velocities for each of the 59 soundings were calculated for the currently reported significant pressure levels and interpolated at 50 mb intervals so that the relative winds for each 50 mb level from the surface to 100 mb could be obtained. Seven storm speeds were used in the calculations: 50, 55, 60, 65, 70, 75, and 80 percent of the mean environmental wind speed; and 26 storm directions were used: from 60 degrees to the left to 60 degrees to the right of the mean environmental wind direction, incremented every 5 degrees. In Figure 9-2 all possible vectors drawn from the center (C) of the cumulus cloud to each intersection of the lines of storm direction and storm speed give the 182 different storm variations used in this study.

The Shear Index

In order to apply realistically the double vortex model to the 59 soundings,

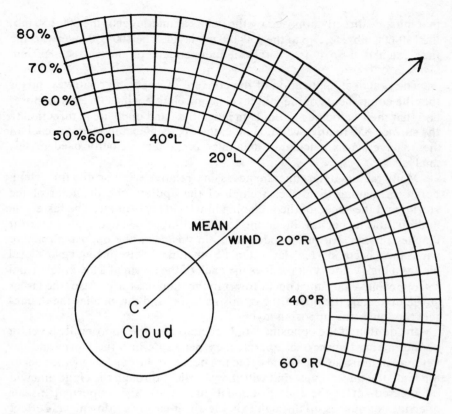

Figure 9-2. Combinations of Storm Speeds (50 to 80% of the Mean Wind) and Storm Directions (60° Right to 60° Left of the Mean Wind) That Were Used for Making Calculations of Relative Winds and Shear Index

a layer vector average of the calculated relative winds was computed. Since the model is based on opposing winds between low and mid-levels, an examination of the relative winds of the mid-level layers to see if they oppose the relative winds of the low level layer should be better than examining each 50 mb level individually. Therefore, 13 layer averages were computed, each composed of four consecutive levels. Because each 50 mb layer contains the same mass of air as any other 50 mb layer, averaging four consecutive 50 mb level relative wind velocities would give the average relative wind vector velocity of a 150 mb deep layer. This thickness of air should be sufficient to successfully block the incoming layer of air causing double vortex thunderstorm development. Average velocities were calculated for 13 layers: The surface-850 mb (used as the low level inflow layer that must be directly opposed by the winds at mid-levels); the 800-650 mb, the 750-600 mb, etc., up to the 250-100 mb layer. The 800-650 layer was the first calculated so it is, therefore, independent of the surface-850 mb average.

The practical application of the concept of opposing relative winds is that the components of the mid-level layer relative winds directly oppose, and that their velocities be equal in magnitude and opposite in direction to the surface-850 relative wind. Hence, the opposing component for each of the 13 layers was calculated and this component velocity used in this analysis.

Many factors determine the necessary relative wind profile for a thunderstorm, among them the strength of the updraft, the diameter of the storm, and the speed of the incoming low level layer of air. The faster the updraft, the higher the incoming low level layer of air travels before it reaches the back of the thunderstorm; in this case, the opposing relative winds must occur in a high layer. On the other hand, if the incoming low level air has a high velocity, it reaches the back of the storm at a lower level and the opposing winds must occur lower in the thunderstorm. Also, the larger storm diameter, the higher the opposing layer, and the smaller the diameter, the lower the opposing layer.

In addition, if the opposing winds occur below the appropriate level for counteracting the thermal updraft they penetrate into the storm and development into a severe storm could not occur. If the opposing winds occur too high, the lower level wind will shoot straight through the cloud and will be sheared off at the top. Frictional effects are also important; as air overrides or pushes up through existing air, it is slowed down. This effect also helps determine where the opposing level must be. Hence, the level of opposing relative wind velocities necessary for the creation of double vortices in a cumulus cloud is quite critical and is different with each storm.

All these effects must be indirectly taken into account in a good forecast index, but all are too complicated to be specifically included for a general forecast. It became necessary, therefore, to set up boundaries for the occurrence of the opposing component velocities. Six layers were chosen as the critical mid-level layers for the occurrence of opposing component velocities; these were the 650-500 mb, the 600-450 mb, the 550-400 mb, the 500-350 mb, the 450-300 mb, and the 400-250 mb layers. For slow updrafts and fast indrafts, the 650-500 mb layer would be expected to oppose the wind; for fast updrafts and slow indrafts, the 400-250 mb layer would be expected to oppose the incoming air.

It is quite possible for the component velocities in more than one consecutive layer to oppose the surface-850 vector wind. Clearly, the more consecutive layers that oppose the incoming air, the deeper and more effective the barrier, and the more likely the winds would collide with the thermal updraft creating a double vortex thunderstorm. Ideally, all six layers would oppose the incoming air, offering an excellent barrier to maintain the rotary motion of the winds.

To allow for all the factors that may vary the speed of the thermal updraft, the opposing components in the six mid-level layers were considered effective if they were within a certain fraction of the surface-850 vector wind. It was assumed that the magnitude of the mid-level components was effective if the value was between 75 percent and 125 percent of the magnitude of the surface-850 wind.

The Shear Index (SI) of a trial storm speed and direction was defined, therefore, as the number of consecutive mid-level layers whose opposing components were between 75 percent and 125 percent of the surface-850 wind speed; the values of the Shear Index, therefore, can vary from zero to six. If an opposing component wind greater in magnitude than the surface-850 wind occurred below the 650-500 mb layer, the thunderstorm could not build and that storm speed and direction combination corresponds to an SI of zero. Also, if the opposing component winds occurred above the 400-250 mb level, an SI of zero resulted. The Shear Index of a sounding was defined as the largest SI for all 182 trials obtained for the seven storm speeds and 26 directions.

An example of a trial storm speed and direction whose SI is six is given in Figure 9-3. Below the 650-500 mb layer the opposing component velocity is less than 75 percent; but at the 650-500 mb layer, it exceeds 75 percent. The barrier to the winds, therefore, does not start too low. The opposing component continues to increase but never exceeds 125 percent of the incoming wind; hence, it never becomes too large to completely cut off the updraft. Because the six consecutive layer component velocities are all within the limits of 75 percent to 125 percent of the inflow speed, the probability that the updraft will be directly opposed in the mid-levels is quite high. Hence, if the thermodynamic parameters of the atmosphere are favorable in the area of this sounding, then it is quite likely that a severe thunderstorm will occur.

The SIs were calculated for all 59 soundings by computer and are listed in Table 9-1; a few examples of how the SI varies with different storm speeds and directions are given in Figure 9-4. The major tornado day proximity soundings of April 11, 1965 (Number 28), June 11, 1970 (Number 18), and June 10, 1967 (Number 1) are shown. It can be seen from this figure that usually less than half of the possibilities gave the best SI for the sounding. Indeed, the average number of trials for the proximity soundings that gave shear indices of zero was 136, or 75 percent, and the average number that gave shear indices of the same value as shown in Table 9-1 was nine, or 5 percent.

A general pattern of SIs can be seen from these diagrams. In all cases for

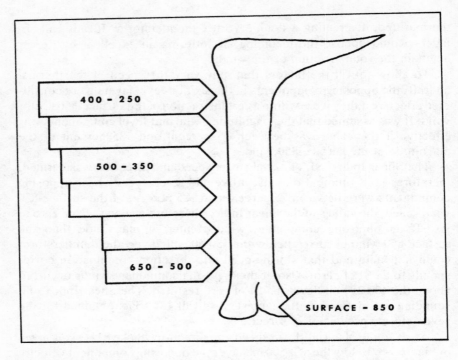

Figure 9-3. An Example of a Maximum Shear Index of 6. Six layers between 650 and 250 mb were within 75 to 125 percent of the surface to 850 mb thermal updraft.

each trial storm speed there is a grouping of storm directions that gives a positive SI, the SIs within that group increasing in pyramid fashion to the best SI for the sounding; the number of possible combinations that give SIs varies from only 18 for Oklahoma City to 81 for Flint Michigan. Soundings that have only a small number of possible storm directions that give positive SIs are from storms that require a critical, well defined storm path in order for the storm to reach tornadic intensity. Soundings that have a larger number of possible storm directions, on the other hand, indicate that the paths for these storms were not as critical and can, in fact, vary as much as 25 degrees. In all cases, the SI for the sounding occurs in a small number of trials, narrowing down the best relative wind profile to a very small number of possibilities. Figure 9-4 also shows that the storm direction trials that gave the best SI for each of the sounding can vary from 35° to the left to 30° to the right. This is in agreement with observations [34] that tornadoes can occur in both left and right moving storms.

The observed storm direction should correspond to one of the calculated values of storm direction used to determine the best SI for each

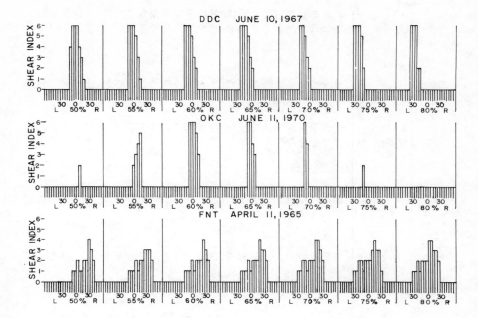

Figure 9-4. Examples of the Variation of the Computed Shear Index for Three Different Tornado Proximity Soundings. The abscissa is the storm direction for various storm speeds.

sounding. To check this the observed motion of the storms for the 27 proximity soundings were correlated with the storm directions corresponding to the best SI. Five soundings either gave an SI of zero or had no observed storm directions listed in *Storm Data*. The results of the 22 soundings for which both calculated and observed data existed are shown in Figure 9-5. The regression coefficient calculated on the computer for these data is 0.94, implying a good relationship. It can be seen from Figure 9-5 that when the observed and calculated storm directions are far apart, the observed storm direction tends to be slightly to the right except for one major anomaly. In other words, these storms may have formed while moving in directions close to those calculated and it was not until they had slowed and veered to the right that the tornado dropped from the cloud. The average observed storm direction for the 22 soundings was 240° and the average calculated was 238°, implying that these severe thunderstorms, on the average, did take advantage of the best wind profile before the tornado dropped from the cloud. The Shear Index, then, appears to be a good measure of the necessary relative wind profile of an air mass that may spawn a tornado.

In order to compare the Shear Index with other indices, the Tornado

Table 9-1

Atmospheric Soundings Used in This Study

No.	Location		Time of Sounding	Time of Tornado	Direction from Station	Distance from Station (miles)	TLI	EI	SI	ESI
1	DDC	6-10-67	1715	1750	NW	20	59.6	-6.37	6	-14.74
2	OMA	7-01-65	1716	1715	SSW	40	39.7	-4.63	6	-11.26
3	DDC	6-13-70	1715	1815	NE	40	70.8	-4.60	5	-10.20
4	TOP	8-06-62	1730	1740	N	25	48.7	-6.04	2	-10.08
5	STC	6-13-68	1715	1850	SW	110	30.6	-3.48	6	-8.95
6	TOP	6-08-66	1715	1900	SW	20	29.6	-3.54	4	-7.07
7	FTW	3-12-71	1726	1730	NE	60	0.0	-2.94	5	-6.87
8	OMA	6-13-68	1716	1720	N	120	46.5	-3.43	4	-6.87
9	AMA	5-31-68	1715	1725	S	45	16.6	-2.36	6	-6.72
10	OKC	6-11-67	1715	1845	E	40	23.4	-2.62	5	-6.23
11	TOP	5-18-59	1730	1850	N	50	65.8	-3.02	4	-6.04
12	CBI	6-08-66	1715	1900	W	200	37.7	-1.86	6	-5.72
13	MGM	4-11-65		TORNADO > 200 MILES			32.5	-1.75	6	-5.50
14	ALB	8-11-66	1815	1800	SE	65	7.3	-1.71	6	-5.42
15	OKC	4-14-65	1715	1730	S	20	23.5	-2.70	4	-5.40
16	OKC	6-08-66	1715	1730	W	45	21.9	-2.65	4	-5.30
17	PIA	6-16-70	1716	1740	ESE	50	15.9	-2.94	3	-4.87
18	OKC	6-11-70	1715	1730	E	60	38.0	-1.32	6	-4.63
19	OMA	6-24-68	1716	1835	NNE	30	15.7	-2.06	4	-4.12
20	GRB	6-04-66	1830	1900	NE	40	19.9	-1.51	5	-4.03
21	SAN	4-11-65		TORNADO > 200 MILES			22.3	-1.97	3	-2.95
22	GSO	4-11-65		TORNADO > 200 MILES			4.9	-0.96	5	-2.65
23	TOP	6-11-67	1805	1800	N	10	14.4	-3.33	0	-2.65
24	DAY	4-11-65	1830	2120	NNW	70	58.7	-1.30	4	-2.61
25	BNA	4-11-65	1735	1908	SW	145	23.1	-2.14	2	-2.29

#		Date								
26	SHV	1-22-65	0515	0615	SW	10	7.4	0.25	6	-1.51
27	LIT	4-30-70	1715	1745	N	30	30.7	-1.76	2	-1.51
28	FNT	4-11-65	1815	1915	SW	110	17.8	-0.72	4	-1.45
29	FTW	6-08-66	1730	1920	NNW	180	17.1	-0.68	4	-1.36
30	OMA	6-08-66	1716	1900	S	160	0.0	0.34	6	-1.33
31	PIA	7-17-68	1715	1830	WNW	50	14.3	-0.56	4	-1.11
32	PIA	1-24-67	1730	1730	SSW	45	24.6	-1.40	2	-0.79
33	JAN	3-03-66	1800	1600	W	20	16.0	-2.14	0	-0.28
34	LIT	4-03-68	1740	1820	SE	65	31.0	-2.09	0	-0.17
35	PIT	4-11-65		1700	TORNADO > 200 MILES		15.1	-0.21	3	0.56
36	DAY	5-08-69	1830		S	5	13.4	-0.67	2	0.67
37	INL	6-13-68		1700	TORNADO > 200 MILES		17.3	1.46	6	0.92
38	MIA	6-25-68	1815	1725	W	1	1.3	-1.00	1	1.01
39	GRB	4-11-65	1715	1615	WSW	140	0.0	1.86	6	1.71
40	JAN	1-23-69	0515	0525	SW	65	28.2	-1.02	0	1.96
41	FNT	8-09-69			TORNADO > 200 MILES		13.3	0.65	3	2.30
42	LIT	6-08-66			TORNADO > 200 MILES		15.4	1.22	4	2.43
43	FTW	4-25-70	0530	0615	E	60	5.2	2.25	6	2.49
44	DDC	6-08-66	1715	1725	NW	180	19.9	1.95	5	2.90
45	PWM	6-21-65	1821	1700	N	60	2.9	-0.44	2	3.12
46	SSM	4-11-65			TORNADO > 200 MILES		0.0	0.84	6	3.69
47	PIA	6-13-68			TORNADO > 200 MILES		4.2	2.91	3	3.82
48	SHV	1-21-65			TORNADO > 200 MILES		9.3	1.53	2	4.06
49	LBF	5-15-68			TORNADO > 200 MILES		3.3	1.14	3	4.27
50	LKC	1-23-69	0515	0525	NE	200	28.2	1.98	3	4.95
51	BIS	6-13-68			TORNADO > 200 MILES		13.3	2.03	3	5.06
52	AHN	4-18-69			TORNADO > 200 MILES		9.7	1.03	0	6.06
53	AMA	4-11-65			TORNADO > 200 MILES		1.9	2.64	3	6.29
54	ALB	4-11-65			TORNADO > 200 MILES		0.0	3.58	4	7.17
55	PIA	4-21-65	1829	2050	SE	90	5.4	1.84	0	7.67
56	LBF	4-11-65			TORNADO > 200 MILES		0.0	2.37	1	7.73
57	GRB	4-21-67	1715	1855	SE	150	9.1	2.14	0	8.28
58	PIA	4-11-65	1730	1600	NE	100	22.6	2.54	0	9.08
59	BIS	4-11-65			TORNADO > 200 MILES		1.5	3.72	0	11.44

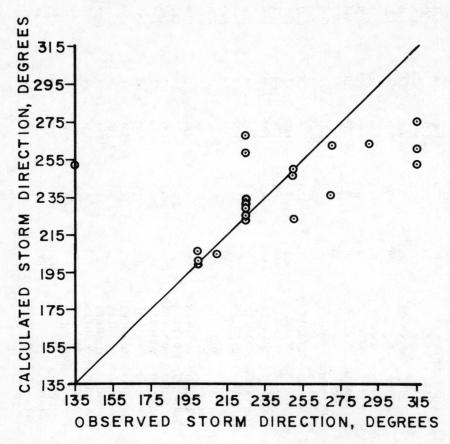

Figure 9-5. Comparison of the Calculated Storm Direction Based on the Best Wind-shear Environment Compared with the Observed Storm Direction

Likelihood Index (TLI) and the Energy Index (EI) discussed previously were calculated for the 59 soundings; the values for these are found in Table 9-1. For a forecast of tornadoes, TLI must be 17 or greater and EI must be −2 or more negative; the forecast ability of each is displayed in Table 9-2. As can be seen from this table, the TLI forecast tornadoes for 18 of 27 proximity soundings and the EI forecast tornadoes for 17 of 27. Out of 20 nonproximity soundings, the TLI forecast no tornadoes 15 times and the EI forecast no tornadoes 19 times. Some of the soundings for which tornadoes were forecast and none occurred were soundings in the warm air ahead of the cold front, but removed from it in both time and distance. This type of sounding has been called a precedence sounding [94]. The TLI forecast

Table 9-2
Summary of the Forecast Ability of the Indexes

	Proximity Soundings for which Tornadoes Were Forecast	Nonproximity Soundings for which Tornadoes Were Not Forecast	Precedence Soundings for which Tornadoes Were Forecast	Precedence Soundings for which Tornadoes Were Not Forecast	Total Correct
Soundings	27	20	12	12	59
TLI	18	15	6	6	46
EI	17	19	1	11	48
ESI	25	18	7	5	55

tornadoes for six of 12 precedence soundings, the EI for one. These forecasts are allowable because the air they were in showed all the characteristics of a tornado-spawning air mass, except that there was no triggering mechanism to start the process; precedence soundings, therefore, can be counted as correct forecasts. This gave a total of 46 of 59 correct forecasts for the TLI and 48 of 59 for the EI; however, the TLI still missed 13 and the EI 11 of the soundings. Hopefully, a better forecast index can be developed that can improve these results. Because the Tornado Likelihood Index does not appear to account for the thermodynamic considerations as well as the Energy Index, the search for a better tornado forecast index was directed toward combining the Shear Index with the Energy Index that is based on thermodynamics.

Energy-shear Index

Because the Shear Index appears to be a good measure of the necessary wind profile for the creation of a tornado producing thunderstorm, when it is combined with the proper thermodynamics such as estimated by the Energy Index, a better tornado forecast index than either index alone should result. In order to discover the best empirical combination of the two indexes, the SI was graphed versus the EI for all 59 soundings; this graph is shown in Figure 9-6. The circled points indicate values from the proximity soundings, the square points indicate values from the proximity soundings for three early morning tornadoes in the Deep South; the x's are values from nonproximity soundings and the triangles are values for the precedence soundings. The proximity and precedence soundings cluster in the lower right-hand corner of the graph where the EI values are negative and the SI values are large. The nonproximity soundings cluster largely where the EI values are positive. A dividing line, therefore, can be drawn

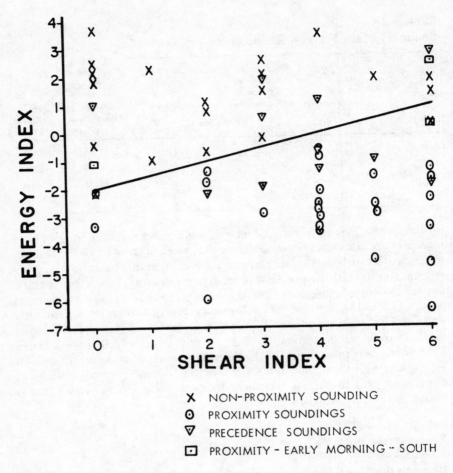

Figure 9-6. A Scatter Diagram of the Energy Index and Shear Index

that separates the proximity and precedence soundings from the nonproximity soundings. Starting at the limiting value of −2 for an SI of zero, a line drawn as in Figure 9-6 separates the nonproximity and proximity-precedence values rather uniquely, although a few discrepancies occur.

There are only two proximity sounding values that lie above the line. Both of these proximity soundings, numbers 40 and 43, are tornadoes that occurred just before dawn in the Deep South. The other tornado proximity sounding that occurred at a similar time in the South is number 26; this sounding is below the line, but close to it. All three soundings had positive EI values, indicating no severe weather was possible; this indicates that the requirements for formation of these tornadoes are different than those for the more usual late afternoon tornadoes. The Gulf Coast or Type II tornado proximity soundings have different characteristics than the usual Great Plains or Type I soundings [86]. It is not unexpected, therefore, that these soundings should be different. It is worth noting, on the other hand, that two of these three proximity soundings had SI values of six.

Seven precedence soundings fall below the line. These were from the June 8, 1966, and April 11, 1965, tornado days; tornadoes probably would have occurred in these areas had there been a cold front and/or divergence and a trough at 500 mb closer to these stations. The two nonproximity soundings whose values fall below the line, numbers 30 and 33, were taken after or near very damaging tornadoes; even though the storms had already passed, the atmospheric conditions that spawned the tornadoes still lingered. Four of the 59 soundings, therefore, are forecast incorrectly—far better results than the TLI or EI gave.

The combined index appears to be a useful forecasting technique for determining from rawinsonde data whether or not tornadoes can occur. The Energy-Shear Index (ESI) can be calculated from the following relationship:

$$ESI = 4 - SI + 2EI \qquad (9.1)$$

If the ESI is negative and a cold front is nearby, tornadoes are predicted; if it is positive, tornadoes are not indicated. Calculated values for each sounding are given in Table 9-1. The performance of the ESI in predicting tornadoes is given in Table 9-2. In 25 of 27 proximity soundings, the ESI forecast tornadoes correctly. In 18 of 20 nonproximity soundings, it correctly indicated no tornadoes. The seven precedence soundings for which tornadoes were forecast were from the major tornado days of June 8, 1966, and April 11, 1965, and, hence, are allowable forecasts. The ESI, therefore, correctly forecast 55 of 59 soundings.

Shear Index Variations with Storm Speed and Direction

The Shear Index has been shown to be a valuable addition to the energy index in calculating the likelihood of tornado occurrences. Because the Shear Index is found by using trial storm velocities chosen within a fixed range of deviations of the mean wind, it should be possible to find the optimum combinations of storm speed and direction that yield positive SI values for the highest percentage of proximity soundings. The frequency of occurrence of a positive SI value for the 182 combinations of storm speeds and directions for the 27 proximity soundings is shown in Figure 9-7.

Storm combinations for which at least 20 percent of the proximity soundings (Figure 9-7) yielded positive SI values ranged between 30° to the

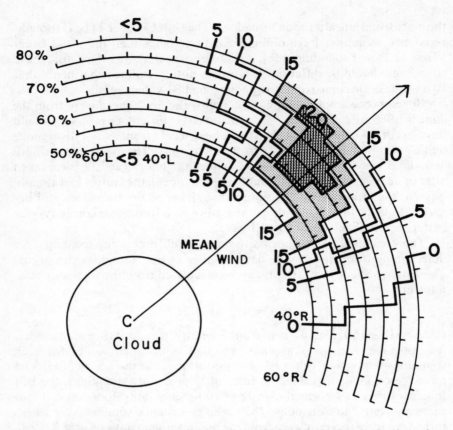

Figure 9-7. The Frequency of Occurrence of Positive Shear Indices for Various Storm Speeds and Directions Calculated from the 27 Proximity Soundings

left and 30° to the right of the mean wind. Deviations greater than 30° to the right or 30° to the left yielded a smaller number of occurrences, most of these with storm directions to the left. Within the first block of combinations, however, an interesting pattern is evident. For the slower storm speeds, deviations of 10° to the right of the mean wind yielded the highest frequency of positive SIs (about 67%), while for the faster storm speeds, 5 and 10 degrees to the left gave the highest frequency (74-81%).

By these empirical results, then, it would appear that potential tornado movement depends on the thunderstorm speed. If it is on the order of 50 to 60 percent of the mean environmental wind speed, probable tornado movement would be 10 to 20° to the right of the mean environmental wind; if it is on the order of 65 to 80 percent of the mean wind, deviations up to 15°

to the left might be expected. The results seem reasonable because it has been observed [34] that many severe thunderstorms split, one part veering to the right and slowing down, the other veering to the left; both are potentially able to produce tornadoes. This type of analysis, then, sheds some light on the expected direction of thunderstorm movement.

In addition, it would appear necessary to calculate SIs for fewer trial storm speed and direction combinations, thus reducing the number of calculations. It appears that deviations ranging from 40° to the left to 40° to the right would be sufficient, thereby reducing the number of combinations from 182 to 119, or 35 percent.

Geographical Testing of the ESI

In order to test the Energy-shear Index on a geographical basis, upper-air sounding data stored on magnetic tape for April, 1970, and June, 1971, were obtained from the National Severe Storms Forecasting Center in Kansas City, Missouri. Maximum tornado days were chosen and the Energy Index (EI) and Energy-shear Index (ESI) were calculated, when possible, by computer from every available upper-air station; the EI and ESI were plotted on separate maps of the United States along with the synoptic situation and confirmed tornado touchdowns four hours either side of the sounding time. Two examples of the major tornado days OOZ June 5, 1971 (16 tornadoes), and OOZ June 10, 1971 (20 tornadoes), are shown in Figures 9-8 and 9-9.

It can be seen from the figures that patterns of Energy Index and Energy-shear Index are roughly the same, with areas of lowest values for each index coinciding; this should be expected because the ESI is based partially upon the EI. In both cases, however, the ESI improved upon the EI; the EI successfully forecast seven major tornado outbreaks in Nebraska on June 5 and 10 in the Texas and Oklahoma panhandles on June 10. In these areas the SI was not quite so critical; a small wind shear would help trigger major storms because of the highly unstable air. However, north of these major tornado outbreaks the wind shear played a much bigger role in causing the other tornado outbreaks. Tornadoes in other words occurred in areas that the EI indicated isolated thunderstorms, but the ESI indicated tornado formation. About half the tornadoes, then, in both cases occurred where the ESI forecast their possibility and the EI did not.

When taken by themselves, both indexes indicate probable tornado formation where none in fact occurred. For instance, both indexes indicate a good chance for tornadoes east of the outbreak areas and along the East Coast on June 5. In both cases these areas lacked a triggering mechanism and the soundings would be classified as precedence soundings. It is

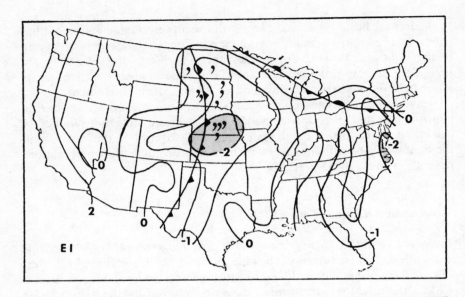

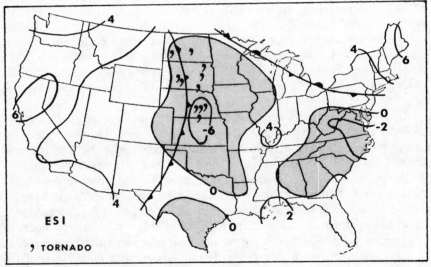

Figure 9-8. The Distribution of the Calculated Energy Index (Upper) and the Calculated Energy-shear Index (Lower) for June 5, 1971

necessary, therefore, to have a triggering mechanism in the area of low index value to forecast tornadoes; on June 5 the mechanism was an advancing cold front combined with a trough, a 50-knot wind core and divergence at 500 mb, and on June 10, a dryline combined with a trough, a 45-knot wind

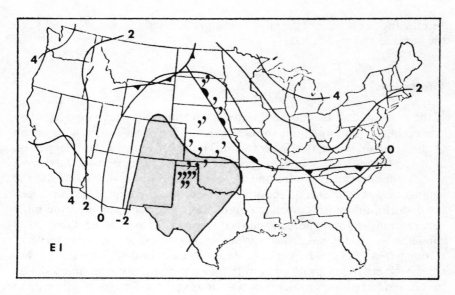

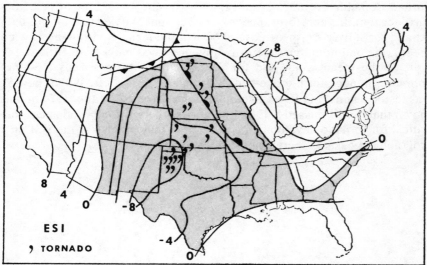

Figure 9-9. The Distribution of the Calculated Energy Index (Upper) and the Calculated Energy-shear Index (Lower) for June 10, 1971

core, and divergence at 500 mb. The areas to the east in both cases were characterized by convergence aloft, which would hinder thunderstorm development. The Energy-shear Index, then, when combined with a triggering mechanism, appears to forecast tornadoes better than does the

Energy Index because of the added parameter, relative wind shear between the surface and mid-levels.

Conclusions

A tornado forecast index, the Energy-shear Index, has been developed by combining a thermodynamic tornado index, the Energy Index, with a wind profile tornado index, the Shear Index. The Shear Index was developed by applying the concept of opposing relative winds as required for the double vortex thunderstorm model and appears to be quite sound. Testing it against observed storm directions revealed that the observed and calculated storm directions correlated quite well. This indicates that the most severe thunderstorms actually moved through the best wind shear environment as calculated from various possibilities of storm movement through the wind field as measured by the rawinsonde. Testing of the Energy-shear Index using proximity and nonproximity soundings shows that it performed better than two other indexes, the Energy Index and the Tornado Likelihood Index. Analyzing the frequency of a positive SI occurrence for different storm speeds and directions showed that for this data set and for slower storm speeds (50-60% of the mean environmental wind speed), the optimum thunderstorm movement was more frequently to the right of the mean wind, and for faster thunderstorm speeds (65-80%) optimum thunderstorm movement was slightly to the left. When the ESI was tested against the EI on two major tornado days in June, 1974, it was shown that the EI missed half the tornadoes where a good wind shear was critical for tornado development. Further testing and use of the ESI on a geographical basis is therefore recommended.

10 Investigation of Wind Pressures on Representative Houses

A systematic evaluation of the damage to houses caused by tornadoes (Chapters 2 and 3) indicates that a severe straight wind is the primary cause of damage to buildings. Houses on either side of the damage path are damaged by winds in the same general direction as the funnel movement at that location. Since straight wind conditions can be modeled in a wind tunnel, an investigation was initiated to determine if the pressure forces from a straight wind would probably produce the type of damage caused by tornadoes. Other objectives of the investigation were to determine the effect of (1) building orientation with respect to the wind, (2) roof angle, (3) boundary layer, (4) appendages; to determine the safest areas in houses when subjected to severe straight winds; and to recommend design parameters and building code criteria to minimize failure under severe straight wind loadings.

An examination of the results of previous tests shows that high values of suction occur near the zones of separation along the sharp edges of the models [95, 96, 97]. These zones obviously are also most critical under a high straight wind condition. Tests by J. O. V. Irminger and C. Nøkkentved [98] and Martin Jensen [99] demonstrate the effect of scaling the wind tunnel boundary layer in model tests. However, in the full-scale problem of a tornado vortex, the boundary layer problem is not a steady flow phenomenon. It more nearly approaches the boundary layer growth problem in a shock tube. Thus, no attempt was made to provide a scaled roughness parameter to produce a specific boundary layer profile. Tests by V. D. Haddon [100] indicated that the values of Ning Chien, et al. [96] and C. A. Salter [97] were unrealistically high. Extensive tests reported by American Society of Civil Engineers (ASCE) [101] considered all possible variations in the horizontal angle. Midwest Research Institute (MRI) [102] and P. D. Gibbs [103] conducted tests on a flat roof building that indicated no apparent increase in suction due to shifting winds. These tests did show that there was one horizontal wind angle that produces a maximum suction on a particular roof.

The work of these previous investigations as well as others [104, 105] provided the basis of setting up a test program.

Test Program

A systematic mapping of the external pressures on 12 representative uilding configurations was made. These included two basements, seven two-story and three one-story house configurations at different roof angles. Tests were conducted on each configuration at each 15° angle of wind at 150 miles per hour wind tunnel speed.

Models

Models were constructed to a 1:48 scale representing basic house shapes. The plan form of each was a rectangle 7.5" - 15". The following configurations were tested:

1. Open basement (four walls)
2. Open walk-out basement (three walls)
3. Two-story house with a flat roof
4. Two-story house with 15° roof
5. One-story house with 15° roof
6. Two-story house with 15° roof and center chimney
7. Two-story house with 15° roof and end chimney
8. Two-story house with 15° roof and gable
9. Two-story house with 15° roof and two gables
10. Two-story house with 30° roof
11. One-story house with 30° roof
12. One-story house with 45° roof

The models were mounted on a turntable in the upper wall of the wind tunnel (Figure 10-1).

Instrumentation

Pressure taps in the models consisted of 1/16" ID brass tubing epoxied flush with the model surface. Tygon plastic tubing 1/16" ID was used to connect the tap to a manifold (Figure 10-2). Pressure in the manifold (when connected to a particular tube) was measured by a differential pressure transducer. The amplified output of the transducer was read from a digital volt-meter.

Figure 10-1. Model Mounted in Test Section

Data

Wind Tunnel Corrections. A model in the test section of a wind tunnel causes a blockage of the cross section by its presence and its wake. In order to correct for errors introduced by wake blockage, a dynamic pressure correction should be made for models having frontal area greater than 5 percent of the test section cross section area [106]. Since the maximum blockage induced by any configuration tested was of the order of 0.75 percent, no corrections were applied. The wind tunnel buoyancy correction factor was also negligible. The tunnel velocity deviation from the mean velocity was less than 1 percent.

Pressure Coefficient. All pressure measurements obtained during the tests were reduced to nondimensional pressure coefficient form. The coefficient of pressure is defined as

$$C_{p_o} = \frac{p - p_\infty}{q_\infty} = \frac{p - p_\infty}{(1/2)\rho_\infty V_\infty} \tag{10.1}$$

Applicability of Data to Full-scale Buildings. The model pressure coeffi-

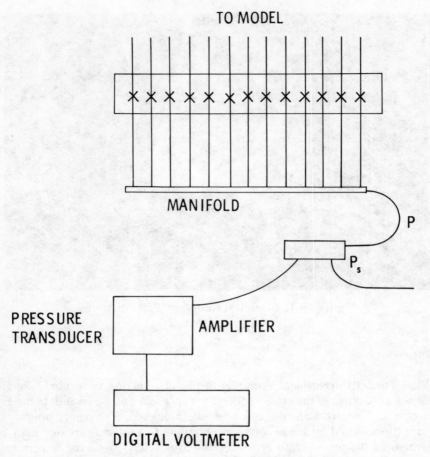

Figure 10-2. Pressure Measuring Instrumentation

cients are applicable for geometrically similar full-scale buildings at the same wind angle when corrected for the effects of a different Mach number, δ/h, K/δ, and R_∞ (Figure 10-3). The coefficient data given are for incompressible flow ($M_\infty \to 0$). These coefficients may be corrected for Mach number by the Glauert equation:

$$C_p = \frac{C_{p_0}}{\sqrt{1 - M_\infty^2}} \tag{10.2}$$

The degree of boundary layer immersion, δ/h, has a significant effect on the magnitude of the pressure coefficients. The roofs of both the one and two story buildings were outside the normal wind tunnel boundary layer. Therefore, the value of wind speed at the actual value of h should be used with these coefficients. However, the destructive wind in the tornado is of

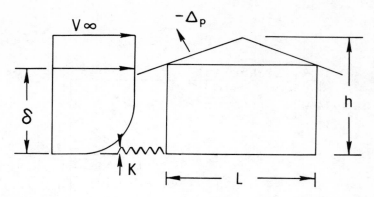

Figure 10-3. The Variables Describing the Functional Relationships for the Pressure Coefficient

such a sudden nature that the boundary layer growth is minimal at the time of the most destructive winds.

The significance of the relative roughness factor, K/δ, depends on the value of δ/h. The effect of a boundary roughness change on the pressure coefficient decreases with increasing roughness and is small with a large roughness factor. The blockage and venturi effects from nearby structures, trees, etc. are probably much more significant. In the absence of data on these latter similarity problems, the roughness factor should be neglected.

In general, the models were blunt sharp-edged bodies. For such shapes with turbulent boundary layers and flow separations, the flow patterns are established by the sharp edges and the influence of Reynold's number on the coefficients is small. The test values of R_n varied from 1.61 to 8.05 $\times$ 10^6.

Pressure Coefficient Data

Pressure coefficient contours for the most critical wind angle of the basic configurations tested are presented in Figures 10-4 through 10-7. Regions in which the pressure coefficients are negative indicate pressures below the atmospheric pressure and local velocity greater than the free stream volocity. When the pressure coefficient is positive, the pressure is above atmospheric and the velocity is less than the free stream velocity. By integrating the pressure coefficients over any complete unit surface area, an average force coefficient may be found for that surface. If this is computed for each wind angle, the maximum force coefficient on the area for any wind angle may be found. Using these values to compute the area forces for a specific wind speed, a minimum structure may then be computed to withstand this wind load.

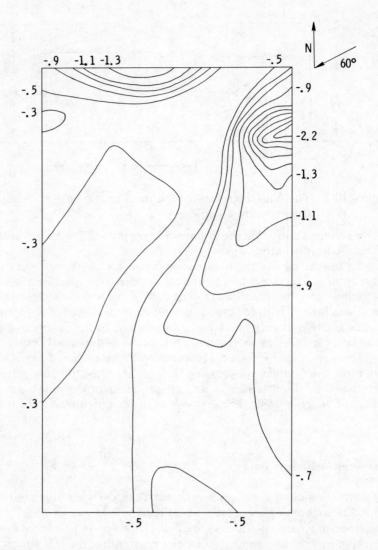

Figure 10-4. Pressure Coefficient Contours for Flat Roof at Critical Wind
Angle of 60°

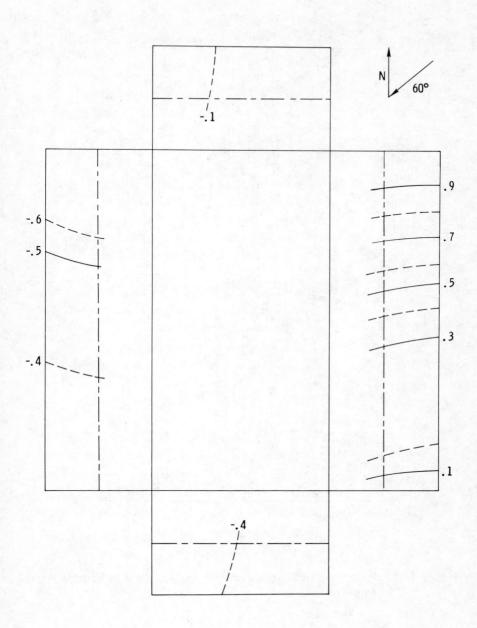

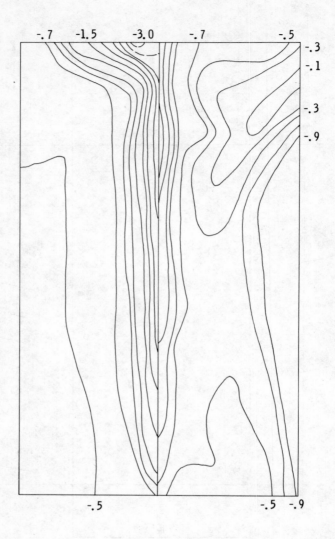

Figure 10-5. Pressure Coefficient Contours for 15° Roof at Critical Wind
Angle 60°

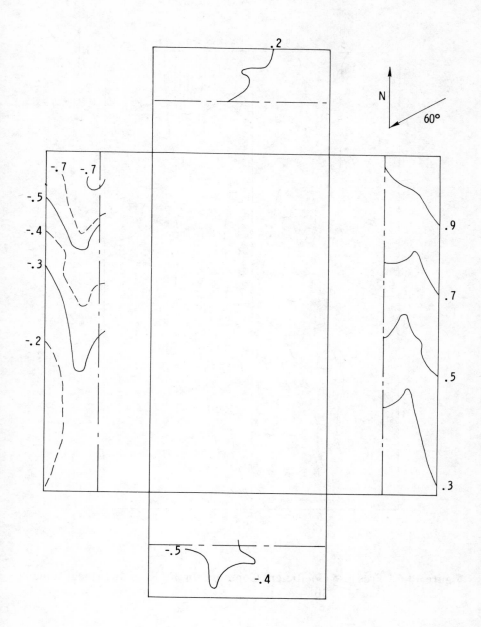

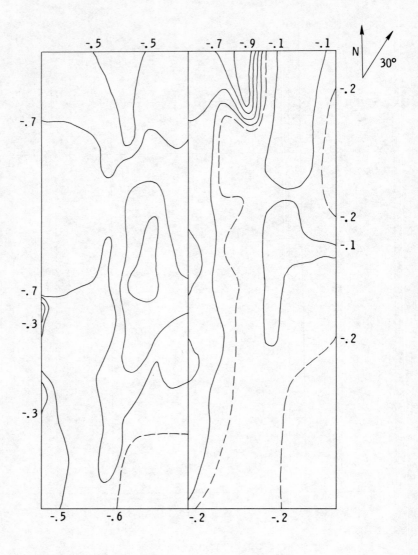

Figure 10-6. Pressure Coefficient Contours for 30° Roof at Critical Wind
Angle of 30°

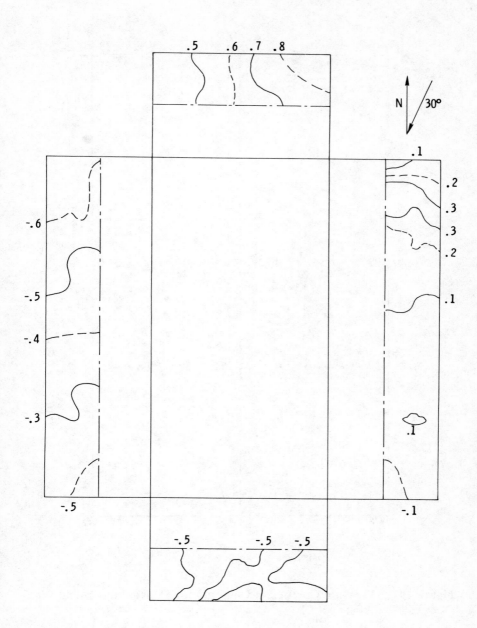

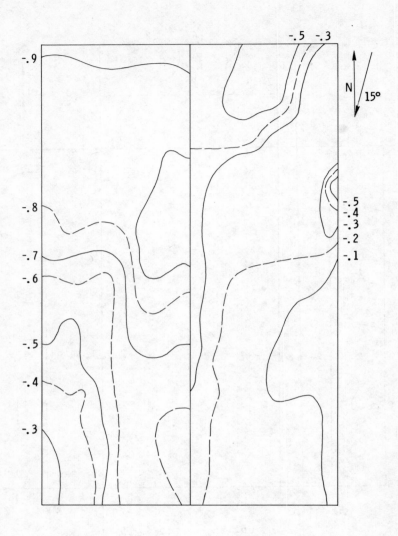

200

Figure 10-7. Pressure Coefficient Contours for 45° Roof at Critical Wind Angle 15°

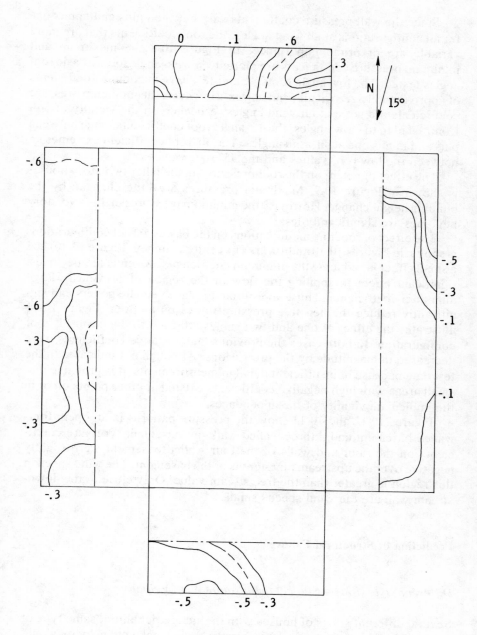

Since the wall pressure coefficients vary between plus and minus one for all configurations at all wind angles, the roof coefficients that are more variable are of prime importance. In Figure 10-8 the maximum and minimum pressure coefficients are plotted vs. wind angle for the basic roof angles tested. The low angle roofs, 0° and 15°, have maximum coefficients of approximately zero at all wind angles. However, the minimum pressure coefficients are large at all wind angles, especially in the vicinity of 60°. Contrasted to the low angles, the 45° angle roof coefficients range between plus and minus one at all wind angles. The 30° roof coefficient extremes lie between the low roof values and the 45° roof values.

The effect of height on the roof pressures in the taller two-story houses is shown in Figure 10-9. Maximum pressures are little changed by the building height change. However, the minimum pressures on the two-story buildings are significantly less.

The effect of roof line modifications on the basic 15° roof configurations is shown in Figures 10-10 and 10-11. The center chimney (Figure 10-10) had a small effect in reducing the minimum coefficients. An end chimney had a significant effect in spoiling the flow in the region of previously large minimum coefficients. Thus, appropriately placed small spoilers can significantly reduce the negative pressure forces on an area. Figure 10-11 illustrates the effect of one and two gables attached to the basic 15° roof configuration. In both cases the previous high negative coefficients were decreased in magnitude by the protrusions. Although not indicated in the test data because of insufficient detail in measurements, it is probable that small areas with high negative coefficients existed at other points on or in the immediate vicinity of the appendages.

Figures 10-12 and 10-13 show the pressure patterns in an open, four-walled basement and a three-walled walk-out basement. Pressure coefficients on the floor and walls of the four walled basement were slightly negative over the upstream nine-tenths of the basement. The wind speed in this region is greater than the free stream value. Only close to the downstream wall are the wind speeds small.

Prediction of Structural Failure

Different Style Houses Subject to the Same Wind Velocity

Several different styles of houses with the same orientation, same type of structural construction, and subject to the same local wind velocity would probably experience different degrees of failures. From Figure 10-8 it is evident that a 15° roof house under most wind angles will be subject to greater roof loadings than the other three configurations. The low roof angles (0°-15°) are particularly more likely to fail than the higher roof angles with wind angles between 45° and 75°. For most wind angles the 45° roof is superior to the other roof angles. Figure 10-9 indicates that the roofs of taller structures will be subjected to slightly larger negative pressures at all wind angles. Thus, for two similar structures, except for height the taller would be expected to fail first.

Roof line changes may reduce negative roof pressure and thus increase the structure's ability to withstand a given wind velocity. Figure 10-10 suggests that appropriately placed spoilers can be used to markedly change the large negative pressures in critical areas. Larger appendages (Figure 10-11) can produce much the same result. Surrounding trees and buildings must also alter the roof pressure patterns. However, it would seem probable venturi effects could also be produced as well as flow attenuation.

Same Style Houses with Different Orientation Subjected to the Same Wind Velocity

Two identical houses oriented differently would generally experience different degrees of failure from the same wind. Figures 10-8, 10-9, 10-10, and 10-11 show that, except for the 45° roof, the minimum pressure is a definite function of wind angle. A 15° roof house is subjected to three times the negative pressure at 60° wind as at a 0° wind angle. Only the 45° roof has small coefficient variation with wind angle.

Initial Mode of Failure of Basic Houses from Pressure Loadings

For the purpose of predicting the initial mode of failure of the basic house, it has been assumed that (1) the roofs of the full-scale structures are uniform in strength and would fail locally when this load is reached; (2) the roof to wall connection may fail before local roof failure; and (3) the wall to foundation connection may fail before (1) or (2) occur.

House with 0° roof. The highest suction pressures occur along the upwind roof edges on the flat roof. The first local roof failure would occur in this region and the downwind region would be the last to fail. If the roof to wall connections were weak, the roof would tend to peel off, hinging about the downwind walls. In both cases the upwind wall would probably be blown inward from the positive outside pressure and negative inside pressure on the wall. The third type of failure that could occur is for the house to be lifted off of its foundation and moved downwind if the roof or roof to wall connections did not fail.

204

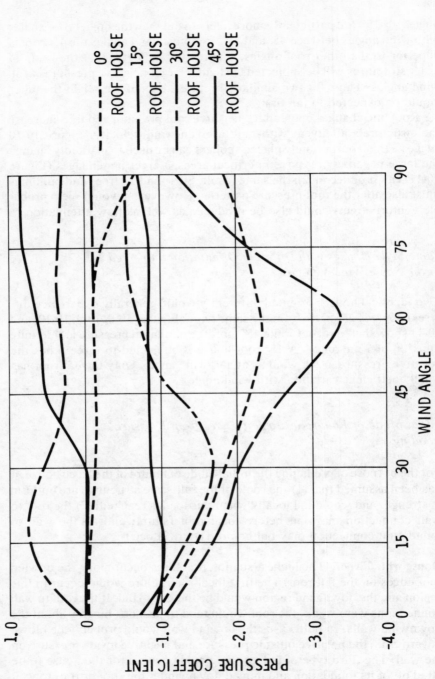

Figure 10-8. Roof Maximum and Minimum Pressure Coefficients

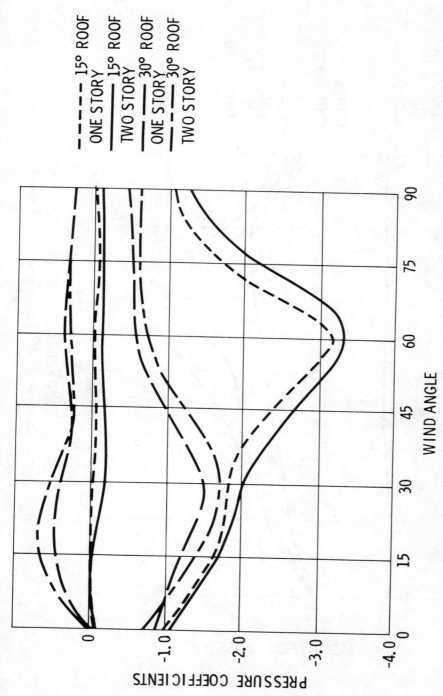

Figure 10-9. Effect of Height on Roof Maximum and Minimum Pressure Coefficients

206

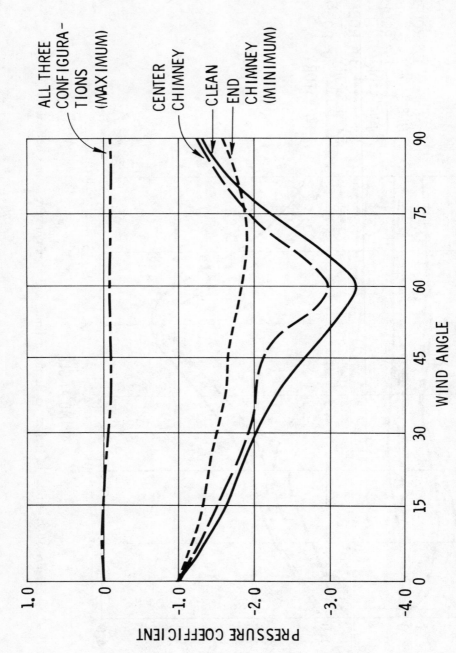

Figure 10-10. Effect of Chimneys on Minimum Pressure Coefficients for 15° Roof

207

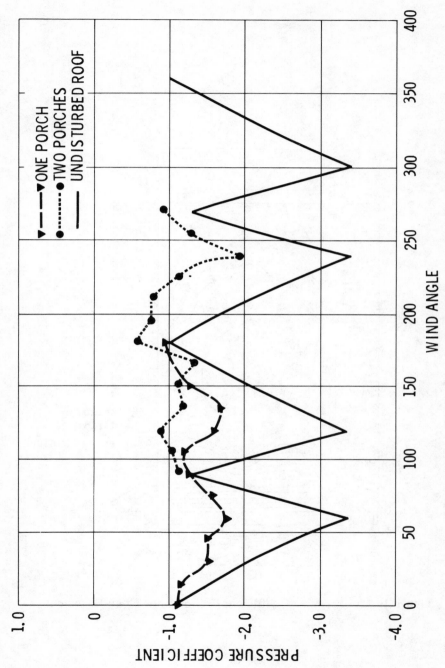

Figure 10-11. Effect of Gables on Minimum Pressure Coefficients for 15° Roof

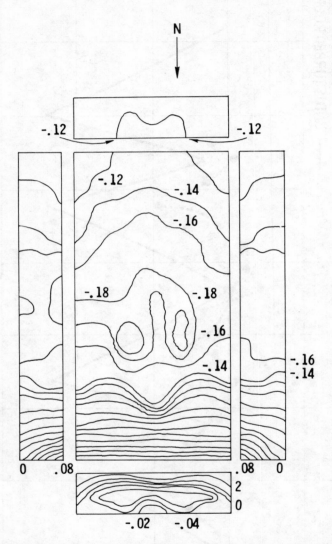

Figure 10-12. Pressure Coefficient Contours for Open Basement

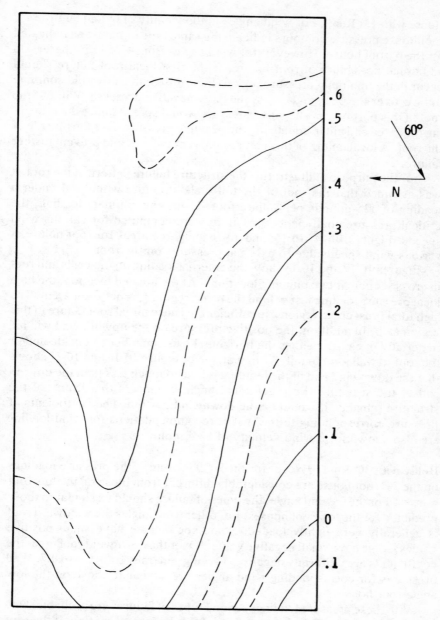

Figure 10-13. Pressure Coefficient Contours for Open Walk-out Basement

House with 15° Roof. For wind angles in the vicinity of 0° and 90°, the 15° roof house pressure contours indicate the same type of failure as indicated for the 0° roof house. However, for wind angles from 45° to 75°, there is an additional possibility of roof failure. If the first local roof failure should occur in the minimum pressure region (Figure 10-5) the air would continue to flow over the peak and create the large negative pressure. Without the roof in this portion, this negative pressure would spread under the remaining roof area and aid in equalizing the negative pressure over the outside of the roof. A local failure of this nature may vent the roof and prevent further failure.

For the purpose of illustrating the structure failure pattern of the roof, it was assumed that the roof of the full-scale structure would fail under a loading of 135 pounds per square foot. Using this ultimate loading, the critical pressure coefficients for failure were computed for various wind speeds. Figures 10-14, 10-15, and 10-16 show the areas for roof failure at various wind speeds due to external pressures on the roof.

Figures 10-17 and 10-18 show the external loading on the walls and roof in cross section at two building locations. At the upwind location, the high local pressure loading just behind the peak, region 1, is quite apparent. The high total pressure in region 2 is sufficient to cause initial roof failure in this region also. In addition, the positive pressure on the upwind wall will aid materially in causing either the horizontal truss member to buckle and/or the roof to wall joint to fail. A comparison of Figure 10-18 and 10-17 shows that the downwind end of the building is loaded much less than the upwind end of the structure. For each wind angle it appears that if any of the structure remains, it would be the downwind portion. The coefficients of pressure corresponding to the ASCE recommendations for wind loading are also shown for a wind velocity of 88.4 mph.

House with 30° Roof. Except for the 90° wind angle, the pressure loadings on the 30° roof house are considerably different from those on the low roof houses. For the 0° wind angle, the types of failure should be similar to those predicted for the low roof houses. For other wind angles the windward roof is generally very lightly loaded. Some coefficients have small positive values; some have small negative values. Over the downwind roof area, the coefficients are uniformly negative, varying generally from 0.4 to 0.7. Wall pressures for corresponding wind angles are similar to those on the low angle roof houses.

With these pressure loadings several types of failures are probable: One type of failure would simply be failure of the downwind roofing. If the roof to wall connection were weak, the whole roof might float off and land a distance downwind of the wall structure. The third type would be for the

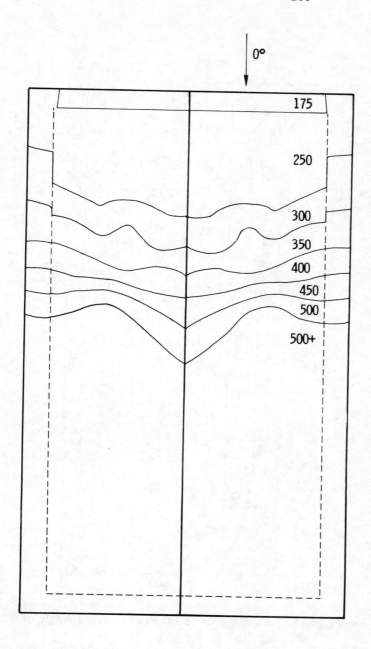

Figure 10-14. Failure Areas of 15° Roof at Wind Angle 0°

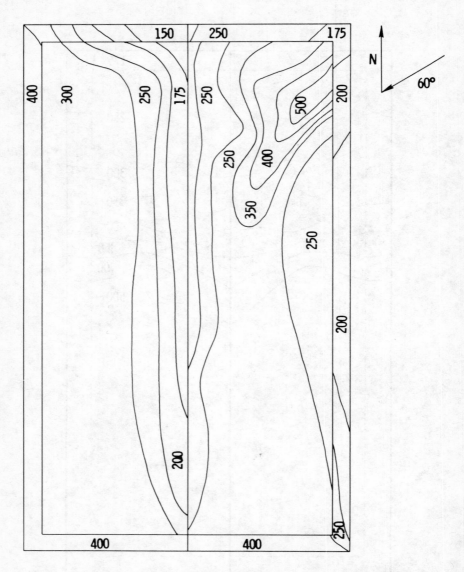

Figure 10-15. Failure Areas of 15° Roof at Wind Angle of 60°

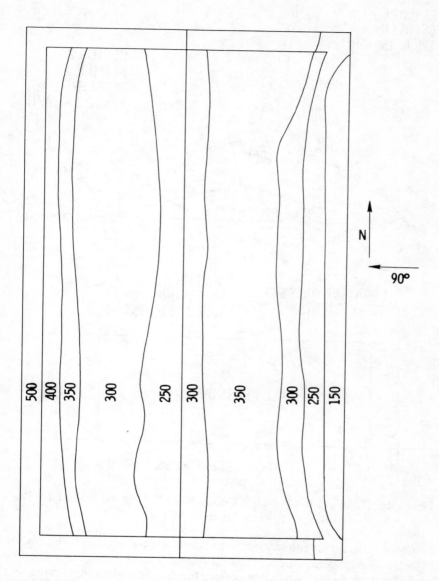

Figure 10-16. Failure Areas of the 15° Roof at Wind Angle of 90°

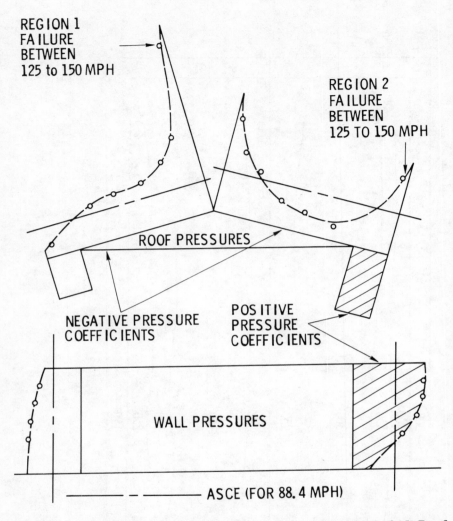

REGION 1
FAILURE
BETWEEN
125 to 150 MPH

REGION 2
FAILURE
BETWEEN
125 TO 150 MPH

ROOF PRESSURES

NEGATIVE PRESSURE
COEFFICIENTS

POSITIVE
PRESSURE
COEFFICIENTS

WALL PRESSURES

ASCE (FOR 88.4 MPH)

Figure 10-17. External Pressure Coefficients at Upwind End of 15° Roof
Building at Wind Angle of 60°

house to slide off its foundation in the downwind direction if the wall to
foundation connection were weak. If the wall structure were weak, the
house might cave in or squash over from the drag force loadings on the roof
and walls. This type of failure would most likely be at wind angles near 30°.

House with 45° Roof. The types of failures that are likely to occur with the
45° roof house appear to be the same as with the 30° roof house. Local roof

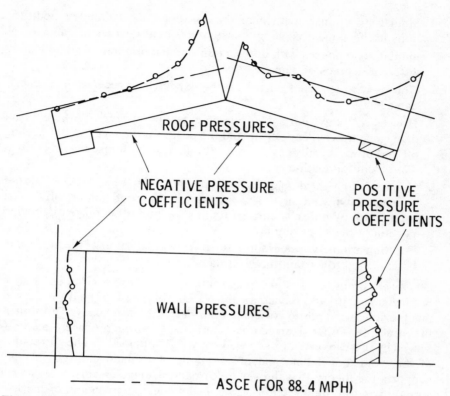

ROOF PRESSURES

NEGATIVE PRESSURE
COEFFICIENTS

POSITIVE
PRESSURE
COEFFICIENTS

WALL PRESSURES

ASCE (FOR 88.4 MPH)

Figure 10-18. External Pressure Coefficients Near Leeward End of 15°
Roof House, Wind Angle 60°

failure would be most likely to occur at wind angles near 15° for the 45° roof
versus 30° wind angles for the 30° roof. The other types of failure indicated
for the 30° roof would most likely occur at wind angles of 45 to 90° for the 45°
roof. In general, the house with a 45° roof appears to be the least susceptible
to failure of the four roofs tested.

*Comparison of Predicted Types of Failures from Wind
Tunnel Pressure Data and Tornado Observations*

The preceeding predictions of structure failure indicate that the following
conditions would be observed in the wake of a tornado if the severe damage
in a tornado is the result of a straight wind:

1. Different style houses in the same area should exhibit different degrees
   of damage.

2. Similar style houses with small dormer, porch, and chimney modifications in the same area should sustain different degrees of damage.
3. Similar style houses with different directional orientation should show different degrees of damage.
4. Houses with low roof angles can be expected to receive the greatest damage.
5. Houses with low roof angles can be expected to receive (1) local roof damage at upwind edge or behind the peak, (b) damage by roof peeling off and upwind walls caving in, or (c) damage by house moving downwind from foundation.
6. Houses with high roof angles may be expected to receive (a) local roof damage downwind of peak, (b) damage by roof sliding off walls, (c) damage by house sliding off foundation and (d) damage by upwind portion of house caving in.
7. For the various types of failure the downwind portion of the house can be expected to sustain the least damage.

Observations of tornado damage indicate that these conditions exist. J. R. Eagleman [16] and Eagleman and V. U. Muirhead [69] have observed that the downwind portion of a house has been the least damaged portion of the structure. Other damage characteristics (1 through 6) were also observed by these investigators as well as by T. T. Fujita [22]. The correlation of the wind tunnel data and observations of tornado damage strongly supports the theory that the severe tornado damage is caused by strong straight wind in the direction of the instantaneous movement of the tornado.

Design Air Loadings

The design air loadings for a building structure should be a composite of the maximums and minimums for the particular roof type and all wind orientations. These loadings are presented in Figures 10-19 through 10-22 for a dynamic pressure of 20 pounds per square foot (a wind speed of 88.4 mph under standard atmospheric conditions) for the four roof angles tested. If it is desired to design the building for higher wind speeds, the values in Figures 10-19 through 10-22 should be recomputed using Equations (6.22) and (6.23).

Chapters 6 and 7 indicate that the winds in the major damage area of a tornado may be of the order of 500 miles per hour. The value of $M_w = 0.756$ (576.3 mph) computed in Chapter 6 will be used to illustrate the recomputation of loadings for higher wind speeds. First, considering a low structure with small δ/h, the effective wind speed would be less than 576.3 mph.

Although there is no method at present to determine the relationship between δ/h and wind speed for this type of flow, assume that for the given δ/h the speed is reduced to 545.0 mph ($M = 0.715$). From Figure 10-22 the loading of -20.9 lbs./sq.ft. was selected as the force, F_1, at the 88.4 mph. From Equation (6.23), F_2 at 545.0 mph may be determined as follows:

$$F_2 = \frac{C_{F_2}S_2}{C_{F_1}S_1} \frac{\sqrt{1 - M_1^2}}{\sqrt{1 - M_2^2}} \frac{q_2}{q_1} F_1$$

Since $C_{F_2}S_1 = C_{F_1}S_1$ and $q = (1/2)\rho V^2$,

$$F_2 = \frac{\sqrt{1 - M_1^2}}{\sqrt{1 - M_2^2}} \frac{\rho_2}{\rho_1} \frac{V_2^2}{V_1^2} F_1 = \frac{\sqrt{1 - 0.116^2}}{\sqrt{1 - 0.715^2}} \frac{\rho_2}{\rho_1} \left(\frac{545}{88.4} \right) F_1$$

$$F_2 = 53.997 \frac{\rho_2}{\rho_1} F_1$$

If it is assumed that ρ_2 and ρ_1 are the densities at two points within a vortex flow at these Mach numbers, then, $\rho_2/\rho_1 = 0.7048/0.9906 = 0.7115$, and $F_2 = 38.42F_1 = -802.9$ lbs./sq.ft. Thus, for the 45° roof configuration, the structure should be designed to support a suction loading of 802.9 pounds per square foot without failure.

Conclusions

1. A straight wind will produce the type of damage observed in the wake of a tornado. The damaging wind in the tornado is a straight wind in the direction of instantaneous tornado core movement.
2. Building orientation with respect to the wind for low roof angle buildings is a key factor in the magnitude of the pressure forces on the buildings. The loading of the 15° roof buildings changes by a factor of 3 with orientation.
3. Roof angle is a big factor in the magnitude of pressure forces on a building. The maximum pressure force on a 45° roof is about one third that of the 15° roof for the same wind velocity.
4. Appendages and spoilers can be used to decrease critical pressure forces.
5. The safest area in all of the configurations tested was the downwind area of the house. Within this area smaller rooms with the most solid wall structure should provide the most protection during a tornado.
6. The recommended wind loading, ASCE (1961) [101] is inadequate.

218

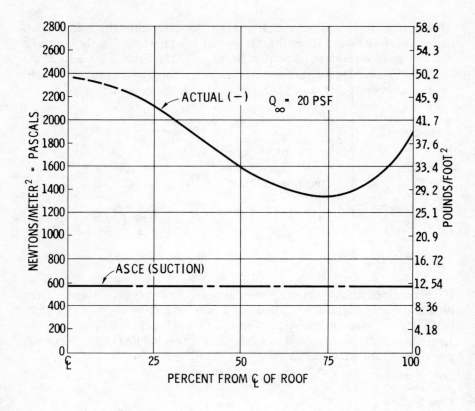

Figure 10-19. Composite Maximum and Minimum Loads on a House with Flat Roof. Roof (upper left), Transverse Walls (upper right), and Longitudinal Walls (lower right).

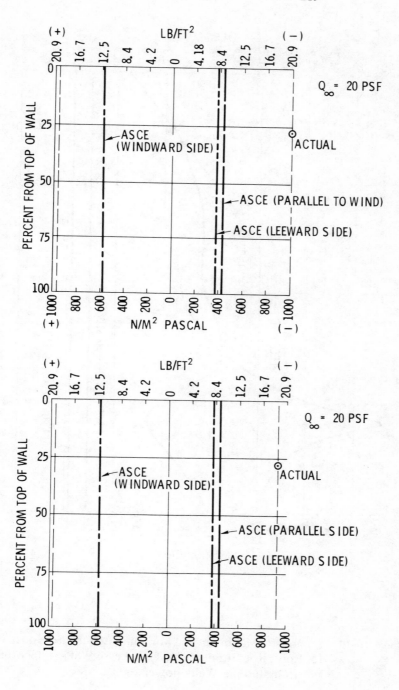

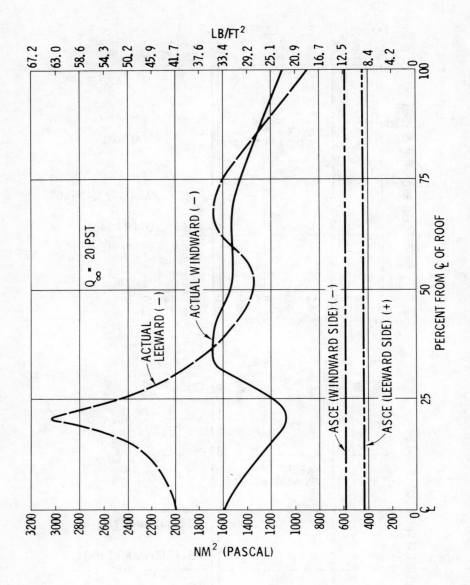

Figure 10-20. Composite Maximum and Minimum Loads on a House with 15° Roof. Roof (upper left), Transverse Walls (upper right), and Longitudinal Walls (lower right).

221

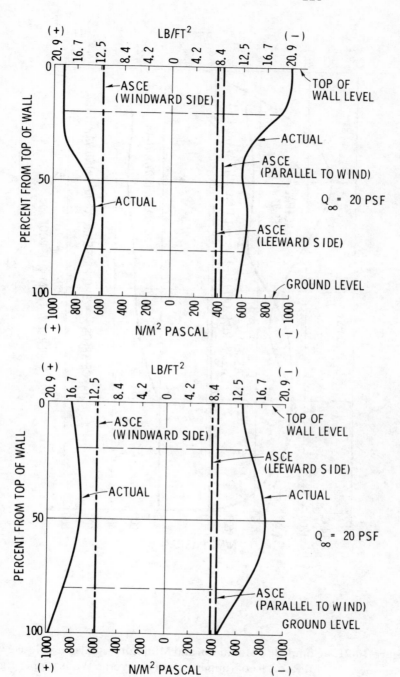

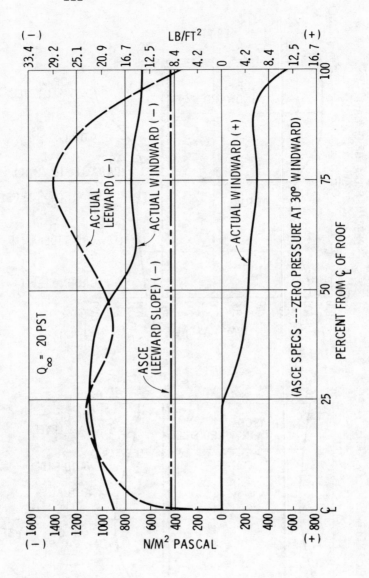

Figure 10-21. Composite Maximum and Minimum Loads on a House with 30° Roof. Roof (upper left), Transverse Walls (upper right), and Longitudinal Walls (lower right).

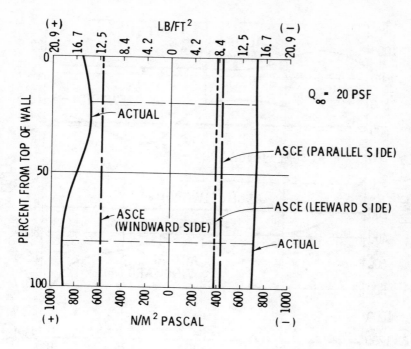

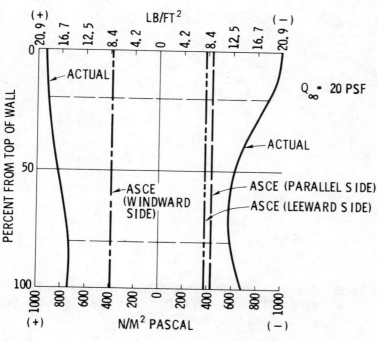

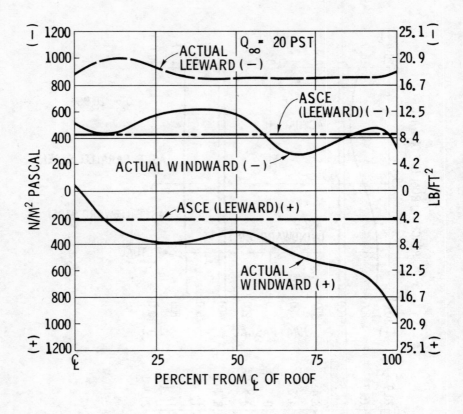

Figure 10-22. Composite Maximum and Minimum Loads on a House with 45° Roof. Roof (upper left), Transverse Walls (upper right), and Longitudinal Walls (lower right).

225

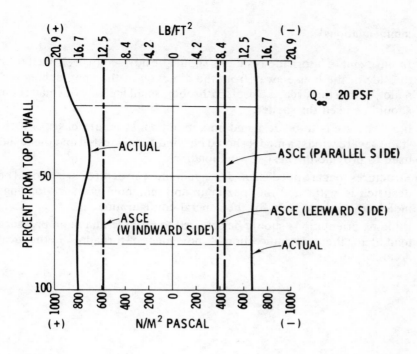

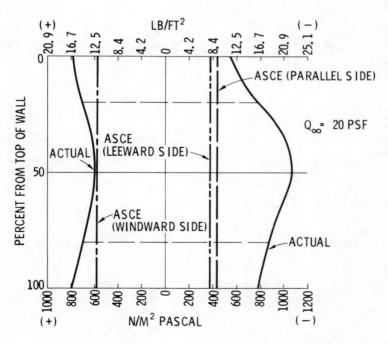

Recommendations

1. In the event of a tornado residents should be advised to seek shelter on the side of the house away from the direction of the approaching tornado. Within this area, a closet, bathroom, small hallway or small room should be used for shelter.
2. If a structure is to be designed on a minimum structure basis to withstand severe winds, a model should be wind-tunnel tested and the result used to provide the design wind loadings.
3. Structures that are not designed on a minimum structure basis should be designed to withstand the maximum and minimum pressure loadings indicated by these tests for the general configuration.
4. Building orientations should be selected to provide minimum pressure loading for the normal direction of movement of tornadoes in the local area.

11

Wind Tunnel Destruction Model Testing

A series of destruction tests were conducted on balsa model houses in the University of Kansas wind tunnel to investigate the modes of failure due to wind pressures. The house configuration chosen for the tests was a rectangular house with a 15° roof positioned in the tunnel to provide a 60° wind angle to the long side of the house. This configuration had the highest roof pressure loading of any configuration tested in a systematic mapping of external pressure on various types of houses. Three types of structural weaknesses were simulated as a result of observed types of failure and of structural design practices and tests made on representative components.

Test Program

Structural Design of Model Houses

Nine models were constructed of balsa. Based on the wind tunnel pressure tests discussed in the previous chapter, it was decided that for destruction, the most critical configuration would be a 15° slope roof with three-foot eaves covering a two-story building 60′ × 24′ (Figures 11-1, 11-2, and 11-3). In view of observed damage of existing houses and the analysis of typical roof structures, it was decided to build a model as true to scale as possible complete with interior partitions, stud-walls, and openings. For simplicity openings were provided over the full height and no distinction was made between the first and second story and no floor was provided.

   In order to simulate the various failure modes, the following variations were considered:

1. Heavy wall to base and roof connections and weak truss connections (using tiny pins as connectors)
2. Strong truss connections and strong sheathing to roof connections, but weak roof to wall connections
3. Strong roof and roof to wall connections, but weak wall to base connections

227

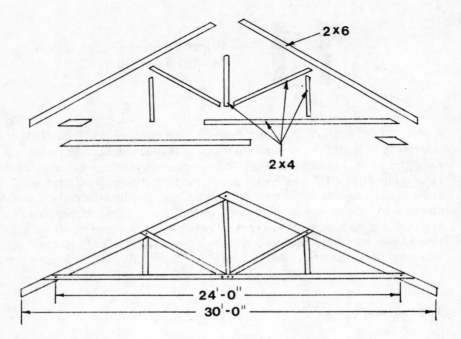

Figure 11-1. Individual Members of the Type I Model Truss Using 11 Separate Members and 8 Wire Pins (Upper) and the Completed Truss

It is obvious that it is practically impossible to simulate the correct relative strength of the three critical connecting areas (foundation to wall, wall to roof, roof itself) and it was, therefore, decided to start out with definite weak roof truss connections (Figure 11-3), and, once this mode of failure was simulated, to strengthen the roof truss considerably (Figure 11-1) and determine by testing which failure mode would occur next. The next step, then, would be to eliminate this by strengthening that failure area so as to induce the next type of failure. In addition, it was decided to experiment with various types of venting to improve the overall resistance against failure. Figure 11-4 shows the building sequence and some details of the first model. The models were mounted upside down in the tunnel. Table 11-1 contains the test configurations.

Instrumentation

The destruction tests were photographed by a 16 mm motion picture

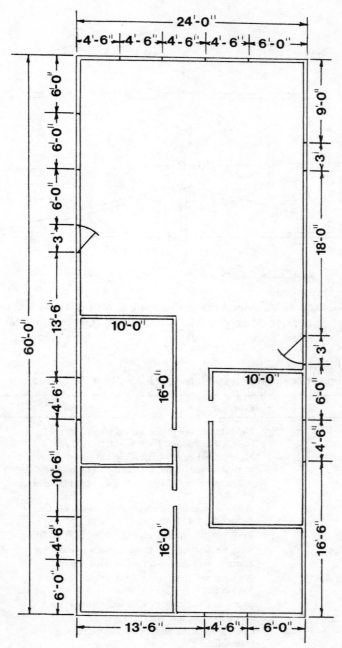

Figure 11-2. Model Plan

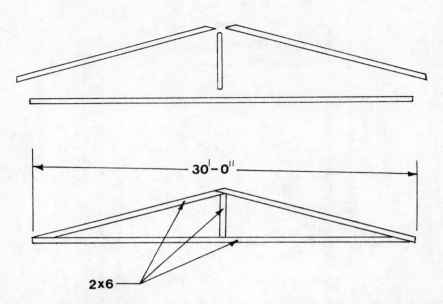

Figure 11-3. Individual Members of the Type II Model Truss Using Four Separate Members and Glued Joints (Upper) and the Completed Truss

Table 11-1
Test Configurations: Two-story House with 15° Roof

No.	Configuration
1	No external openings, no ceiling, weak roof, clamped and taped foundation
2	No external openings, ceiling closed, weak roof to wall connections, clamped and taped foundation
3	No external openings, ceiling closed, clamped foundation
4	Window openings on lee sides, ceiling closed, clamped foundation
5	Roof vented, ceiling vented, clamped foundation
6 (a & b)	Roof vented, ceiling vented, glued foundation
6 (c)	Same as 6 (a & b) with lee window open
7 (a, b & c)	No external openings, ceiling vented, glued foundation
7 (d)	Same as 7 (a, b, & c) with lee window open
7 (e)	Same as 7 (d) less roof
8 (a)	Open front window, ceiling vented, glued foundation
8 (b)	Same as 8 (d) less roof
9	Roof vented, ceiling vented, glued foundation

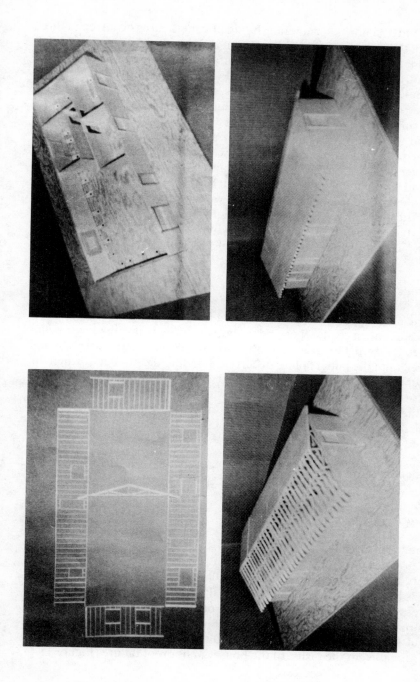

Figure 11-4. Construction of the First Laboratory Model

camera and a 35 mm still camera. The significant failures occurred in three frames of the 16 mm motion picture film at a speed of 64 frames per second. Unfortunately, because of the uncertainty of the time of failure, photographs of the failures were not obtained on a number of tests.

Commencing with model 4, pressures were measured inside the model at two locations. One measurement was made in the area between the roof and the ceiling. The second measurement was made in the area between the floor and the ceiling. Although the latter area was partitioned, the doors were open. The pressure measurements were made with Statham pressure transducers. The amplified transducer output was read by a digital voltmeter and automatically recorded on punch tape by the analog-digital data system of the wind tunnel. The tape data were next converted to pressures and coefficients of pressures by computer.

The dynamic pressure was measured by the analog-digital data system of the wind tunnel and by observation of an alcohol slant manometer.

Test Procedure

After the model was mounted in the wind tunnel and the data system and camera equipment ready, the wind tunnel was turned on with the propeller pitch set for zero tunnel air speed. The tunnel air speed was gradually increased until the initial failure occurred. If failure did not occur when the maximum tunnel speed was reached, the tunnel was shut down and a modification made to the model. A new test was then initiated. In cases where the initial failure was of a minor nature, the test was resumed after an examination of the failure.

Discussion of Results

Configuration 1

The first test model was constructed to test the effect of the pressure loading on the roof and eaves. There were no openings in the model. Pressure tests as described in Chapter 10 indicate two failure areas. One region is behind the peak of the roof, the other, the windward roof corner. The pressure study indicates if any of the roof is left it should be the leeward roof corner. Figure 11-5 shows the initial failure behind the peak of the roof at about 125 miles per hour. This allowed the below atmospheric pressure in this area to vent the inside of the house. Major roof failure (Figure 11-6)

Figure 11-5. Initial Roof Failure at 125 mph—Configuration 1

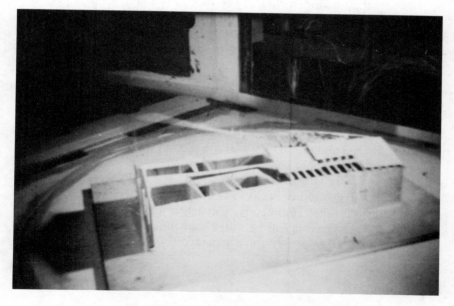

Figure 11-6. Major Roof Failure at 153 mph—Configuration 1

Figure 11-7. Wall Buckling After Roof Failure at 153 mph—Configuration 1

occurred at the upwind corner of the roof at about 153 miles per hour. Figure 11-7 shows the buckling of the upwind wall, partitions, and roof trusses. It will be noted that the least damaged area is the downward part of the house. This test confirms the type of roof failure predicted by the pressure loadings.

Configuration 2

The second test model was similar to the first model except that the roof was strengthened and the roof to wall connection weakened. At a speed of approximately 160 miles per hour the roof "flew off" of the walls. The remaining structure is shown in Figure 11-8. The walls were then tested. The upwind wall was blown back (Figure 11-9). At a speed of over 160 miles per hour the walls failed completely. The final remaining structure is shown in Figure 11-10. Again it will be noted that the downwind corner of the house is the last to be structurally damaged (Figure 11-9).

Configuration 3

Test model 3 was constructed to have its weak point in the foundation to

Figure 11-8. Roof Failure at 160 mph—Configuration 2

Figure 11-9. Wall Buckling at Over 160 mph—Configuration 2

Figure 11-10. Wall Failure—Configuration 2

wall connection. The initial failure was a small lengthwise crack in the balsa roofing behind the peak of the roof. The tunnel speed at this failure was not recorded, but the speed was increased up to 195 miles per hour when total failure occurred. Figure 11-11 shows the building intact in the upper frame. The next frame below shows the house being torn from its foundation. The house was rotated about the rear foundation. In the lower frame the house has been blown completely from its foundation. If the balsa model had had sufficient weight and had been mounted right side up in the tunnel, the house may have slid off the foundation rather than rotated.

Configuration 4

Model 4 was constructed the same as model 3 except that windows were opened on the leeward side of the model. The model failed very suddenly. The roof left the walls nearly intact; the walls and partitions splintered into a number of parts. The failure occurred at 121 miles per hour. The coefficients of pressure and the pressures in the roof area and the below ceiling area are shown in Figure 11-12. The effect of the open rear window can readily be seen in the lower speed of failure from test 3. Model 3 was vented to a much lower pressure in the roof area than model 4 with the open windows. This lower internal pressure in model 3 would be more effective in balancing the low external pressure on the roof.

Figure 11-11. Foundation Failure at 195 mph—Configuration 3.

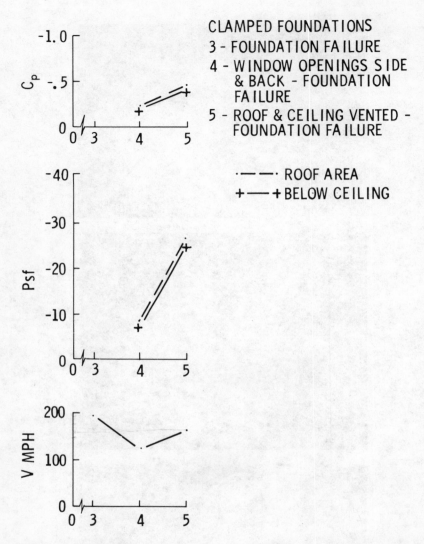

Figure 11-12. Comparison of Test Data—Configurations 3, 4, 5

Configuration 5

Model 5 differed from model 3 in that the roof was vented and small holes were placed in the ceiling to vent the roof and area below the ceiling. Figure 11-13 shows the roof venting. Failure as in model 4 was sudden and in about the same manner. However, as seen in Figure 11-12, it occurred at 159 miles per hour. The internal pressures and coefficients were much higher than those at failure in model 4. Thus, from 3, 4, and 5 it appears that venting the roof to the low pressures downstream of the roof peak provides lower internal pressures and enables the structure to withstand higher air speeds. The opening of leeward windows appears to be undesirable. Self-venting roofs or special roof vents are greatly preferable.

Configuration 6

Model 6 was constructed similarly to models 3, 4, and 5. There were two differences between models 5 and 6. Model 6 was glued to the foundation board. This provided not only greater strength, but with the use of tape, added more strength and sealed the foundation area as well. The roof was vented as in model 5. After two runs, (6a and 6b) at 182 to 182.5 miles per hour, no failure had occurred. A third run (6c) was made with a window cut in the leeward side of the house. Failure of the backside of the roof (Figure 11-14) occurred just below 180 miles per hour. The walls and partitions remained intact during roof failure (Figures 11-15 and 11-16). The plot of internal pressure (Figure 11-17) shows the effect of the open window. Although the roof vent was able to maintain a low pressure under the roof, the open window prevented the pressure under the ceiling from being low also. The net effect was that the roof could not withstand the high speed that it could with the total venting from the roof.

Configuration 7

Model 7 was constructed identical to model 6 without the roof vent. In three successive tests cracks appeared in the roof on the downward side and vented the internal sections of the house. (Cracks were taped after a test and new cracks appeared). On test 7a the maximum speed reached was 192 miles per hour without roof failure; test 7b maximum speed was 180 miles per hour without roof failure; test 7c, 189 miles per hour. Figure 11-18 shows the type of cracks that occurred on the lee side of the roof.

Tests 7d and 7e were conducted after a window was opened on the lee side of the building. Initial roof failure at 141 miles per hour is shown in Figure 11-19. Total roof failure occurred at 180 miles per hour (Figure 11-20). After an inspection of the model, test 7e was run. Wall failure occurred at 188 miles per hour (Figure 11-21).

Pressures and pressure coefficients at failure are shown in Figure 11-22. Tests a, b, and c show nearly constant C_p values as a result of the crack venting in the roof. By opening the lee window, the below ceiling area pressure was much less negative. The roof pressure was also affected by

Figure 11-13. Vented Roof—Configuration 5

Figure 11-14. Initial Roof Failure Below 180 mph—Configuration 6c

Figure 11-15. Final Roof Failure at 183.5 mph—Configuration 6c

the open window. The result was an early roof failure. After this lee roof failure the roof venting increased the negative pressures and total roof failure was delayed. In test 7e the ceiling holes and cracks vented the area below the ceiling. Although the ceiling was very light structurally, it remained intact for some time at tunnel speeds in excess of 180 miles per hour. The walls and partitions failed progressively at 188 miles per hour. Figure 11-21 shows the walls during failure.

Configuration 8

Model 8 was constructed the same as model 7 except with an open wind-

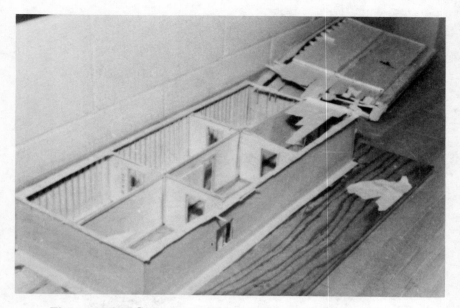

Figure 11-16. Components After Failure—Configuration 6c

ward window. As shown in the lower frame of Figure 11-23, the roof was lifted and rotated from the walls leaving the ceiling. This occurred at a speed of 125 miles per hour. After inspection of the model, test 8b was conducted on the wall and ceiling structure. Cracks appeared in the ceiling (Figure 11-24), the ceiling ripped off (Figure 11-25), and the upwind walls began to buckle below 150 miles per hour. At 150 miles per hour the upwind corner began to be blown in (Figure 11-26) and this portion of the walls was carried away below 180 miles per hour (Figure 11-27). Figure 11-28 shows the last portion of the structure to remain at a tunnel speed of 182.5 miles per hour. This portion was carried away after a few minutes exposure at this speed, leaving only the foundation.

The pressure coefficients and internal pressure at failure are shown in Figure 11-29. Test 8a produced a failure at a much lower speed than test 7d (corresponding test with leeward window open). Internal pressures were much less negative. The disastrous effect of open windward window is obvious.

Configuration 9

The model was constructed nearly identical to model 6. The roof and ceiling failed at 186 miles per hour. The pressure measurements and coefficients at

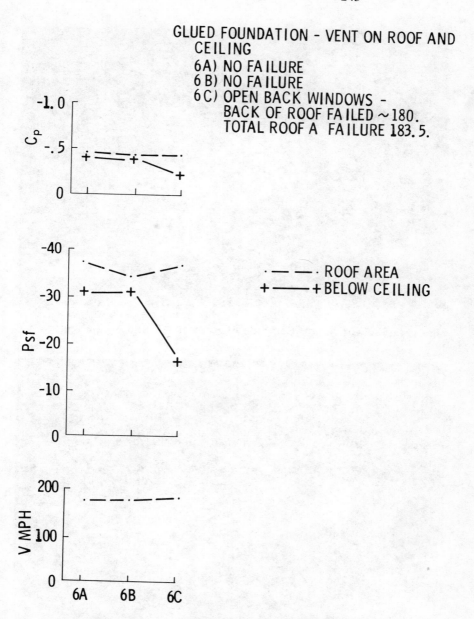

GLUED FOUNDATION - VENT ON ROOF AND
CEILING
6A) NO FAILURE
6 B) NO FAILURE
6 C) OPEN BACK WINDOWS -
BACK OF ROOF FAILED ~180.
TOTAL ROOF A FAILURE 183.5.

· — — · ROOF AREA
+ —— + BELOW CEILING

Figure 11-17. Comparison of Test Data—Configuration 6

failure were close to those for model 6 in tests a and b (Figures 11-17 and
11-29).

Figure 11-18. Cracks on Lee Side of Roof at 192 mph—Configuration 7

Figure 11-19. Partial Roof Failure at 141 mph—Configuration 7d

Figure 11-20. Complete Roof Failure at 180 mph—Configuration 7d

Figure 11-21. Wall Failure at 188 mph—Configuration 7e

REMARKS: OPEN REAR WINDOWS ALLOWED LESS
NEGATIVE PRESSURE IN FIRST FLOOR. EARLY REAR
ROOF FAILURE. RESULTING ROOF VENTING CAPACITY
OVERCOME REAR WINDOW EFFECT BEFORE TOTAL ROOF
FAILURE.

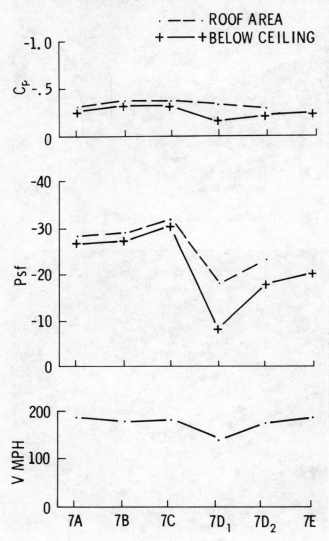

Figure 11-22. Comparison of Test Data—Configuration 7

Figure 11-23. Roof Failure at 125 mph—Configuration 8

Summary of Results

A summary of the test results is contained in Table 11-2. Considering the problem of consistent structural modeling and construction, the test results for the same test conditions check out well qualitatively. The advantageous effect of proper roof venting either by planned means or local roof failure was evident throughout the test program. The desirability of having all exterior wall window and doors closed was also demonstrated. The failure modes were those expected from pressure analysis and model structural changes.

248

Figure 11-24. Ceiling Cracks Below 150 mph—Configuration 8b

Figure 11-25. Ceiling Failure Below 150 mph—Configuration 8b

Figure 11-26. Wall Buckling at Approximately 150 mph—Configuration 8b

Figure 11-27. Upwind Wall Failure Below 180 mph—Configuration 8b

Figure 11-28. Progressive Wall Failure at 182.5 mph—Configuration 8b

Conclusions

1. A 15° roof house presents the greatest resistance to damage from high winds when all outside doors and windows are closed rather than open.
2. The resistance to damage of the 15° roof house is significantly improved by the proper roof and ceiling venting on the leeward side of the roof.
3. Damage modes can be predicted from an external pressure survey of a house constructed by use of standard design practices.
4. The wall to foundation and wall to roof connections should be equivalent in strength to the truss and general structural design strength.
5. If any part of the structure remains, it will most likely be the leeward corner of the house with the 15° roof.

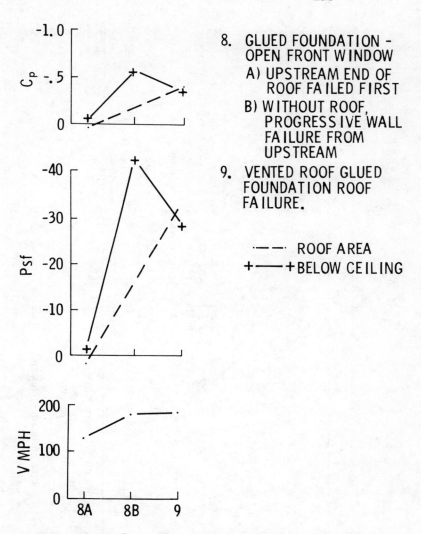

8. GLUED FOUNDATION -
   OPEN FRONT WINDOW
   A) UPSTREAM END OF
      ROOF FAILED FIRST
   B) WITHOUT ROOF,
      PROGRESSIVE WALL
      FAILURE FROM
      UPSTREAM

9. VENTED ROOF GLUED
   FOUNDATION ROOF
   FAILURE.

   ·—·—· ROOF AREA
   +———+ BELOW CEILING

Figure 11-29. Comparison of Data—Configurations 8 and 9

Table 11-2
Summary of Test Results

Configuration	q_∞	V_{mph}	P_{Rpsf}	P_{1psf}	C_{P_R}	C_{P_1}	Remarks
1	—	153	—	—	—	—	Roof failure
2	—	(160)	—	—	—	—	Roof failure followed by wall failure
3	—	195	—	—	—	—	Crack observed in roof before total failure by ripping off foundation
4	36.4	121.5	−8.2	−7.8	−0.225	−0.214	Foundation failure
5	59.2	159.0	−26.1	−25.4	−0.442	−0.429	Foundation failure
6a	70.9	182.0	−37.0	−31.5	−0.465	−0.395	No failure
6b	80.3	182.5	−34.3	−31.4	−0.426	−0.390	No failure
6c	81.0	183.5	−36.0	−16.2	−0.444	−0.200	Roof (rear) failure
7a	87.0	192.0	−28.8	−26.5	−0.332	+0.305	No failure—roof cracks
7b	77.0	180.0	−29.1	−27.0	−0.378	−0.350	No failure—roof cracks
7c	84.7	189.0	−32.6	−30.1	+0.385	−0.356	No failure—roof cracks
7d	48.4	141.0	−18.0	−8.7	−0.372	−0.180	Roof failure
7e	77.3	188.0	−25.2	−20.9	−0.297	−0.248	Wall failure
8a	38.2	125.0	+1.98	−1.14	+0.051	−0.030	Roof failure (upwind)
8b	82.0	182.5	−42.2	−42.2	−0.514	−0.514	Progressive wall failure
9	81.9	186.0	−33.2	−28.1	−0.405	−0.343	Roof and ceiling failed

C_p Coefficient of pressure
P Local static pressure
P_∞ Free stream static pressure
q_∞ Free stream dynamic pressure

All values for pressures and pressure coefficients are extreme values.

$$C_p = \frac{P - P_\infty}{q_\infty}$$

12

Structural Design Practices and Tests on Representative Structural Components

Analysis of Typical Houses

Analysis [107] was conducted on the structural systems of houses with 0°, 15°, 30°, and 45° roof angles using American Society of Civil Engineers (ASCE) specification [101] wind loadings and the maximum loadings developed in the wind tunnel converted to 88.4 mph. The loadings for both ASCE and wind tunnel loadings on the roof trusses are shown in Figures 12-1 to 12-3. These loadings assume a truss spacing at two-foot centers and eaves of two feet. Dead loads were estimated based on standard construction practices and dead load forces were then considered in calculating member forces. A simply supported span and perfectly connected pin supports were assumed. While the joints are possibly more rigid than pin connections, there is no way to evaluate the rigidity provided. However, some of the advantages of rigid joint behavior may exist, especially if gusset-plate connections are used as suggested in Chapter 14.

Combined axial forces and bending moments were considered in the evaluation of the top chords. Both the top and bottom truss chords were assumed to be braced over their entire length by roofing and ceilings, respectively. Thus, the full axial capacity of the members could be developed. The lower of the two members in the upper truss chord, always on the leeward side, generally developed the most critical combination of forces. The axial forces as found in the wind tunnel loadings are as follows:

15° Truss	2988 lbs.	(13,300 N)
30° Truss	331 lbs.	(1,470 N)
45° Truss	95 lbs.	(200 N)

The distributed loads generating the moments, are given in Figures 12-1 to 12-3. As can be seen from the previous data the roof angle is a significant factor in the magnitude of the forces generated. In general the 15° angle caused the greatest forces as is evident upon comparison of the loadings in Figures 12-1 to 12-3.

Evaluation of Present Structural Methods

General

Present commonly used home building techniques (see Figure 12-4) were

253

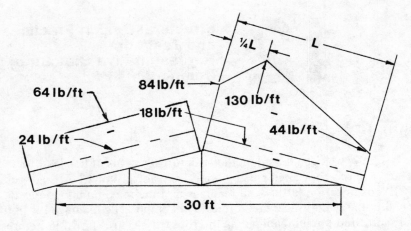

Figure 12-1. Comparison of Wind Tunnel Loadings at 88.4 mph (Solid Line) and ASCE Loadings (Dashed Line) on a 15° Roof with Two-foot Center to Center Truss Spacing and Two-foot Eaves

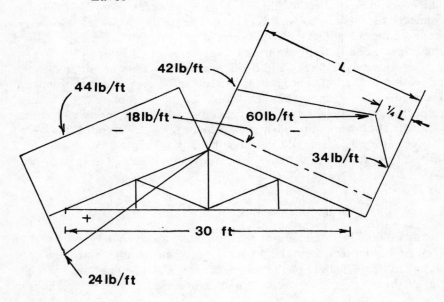

Figure 12-2. Comparison of Wind Tunnel Loadings at 88.4 mph (Solid Line) and ASCE Loadings (Dashed Line) on a 30° Roof with Two-foot Center to Center Truss Spacing and Two-foot Eaves

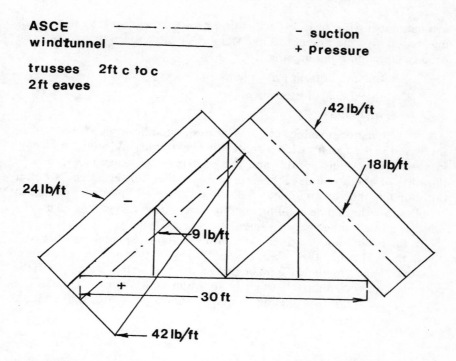

Figure 12-3. Comparison of wind tunnel loadings at 88.4 mph (solid line) and ASCE loadings (dashed line) on a 45° roof with two foot center to center truss spacing and two foot eaves

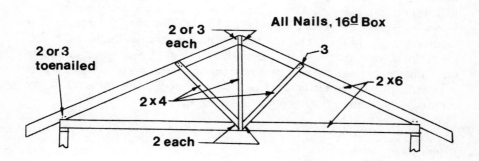

Figure 12-4. Typical Truss Details at Local Subdivision

evaluated using the ASCE specification wind loadings and the wind loadings based on wind tunnel testing. Members were chosen that seem to be used in common building practice;

1. 2″ × 6″ top and bottom truss chords
2. 2″ × 4″ remainder of truss members
3. 2″ × 4″ wall studs.

An average factor of safety of 2.5 was used, and allowable stresses were increased 33 percent for wind loading. Since extreme wind values may only last for a very short period, and the 33 percent increase is based on a duration of about one day, this increase is rather conservative. A better duration might be one minute, which would allow a 70 percent increase in stresses. Southern pine building materials were used and more specifically a No. 2 grade lumber was assumed.

In the evaluation of existing structural methods, special attention was given to evaluating the relative importance of two modes of failure: (1) weak connections and (2) inadequate members. In the evaluation of the possibility of roof failure, particular attention was given to (1) failure of joints, (2) failure of roof sheathing, and (3) failure of roof to wall connections.

Members and Sheathing

The sheathing that is generally used is adequate for the loads generated by the wind tunnel testing. Failure of sheathing is due to poor attachment to the structural system.

In the cases of members themselves, all joints were assumed to be capable of transmitting the various loads. Since the 15° truss was subjected to the greatest loads, failure would first occur in this particular case. However, under the loadings of the 88.4 mph wind tunnel velocities, all members were, in themselves, completely adequate for the loads. It follows, then, that the 30° truss construction would also have adequate strength members. In the case of the 45° truss the strength of the members is adequate, provided that bracing is used so that the full capacity of the internal truss members may be developed.

Connections

The connections used in home building are definitely the weakest link in the structural system. The joints will generally fail well before the adjacent members can develop their capacity. Typical connections were tested at

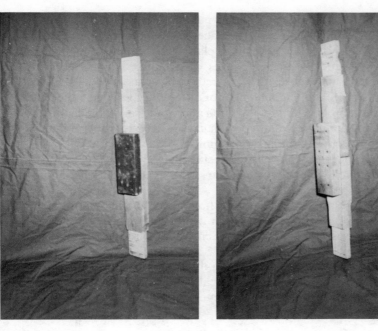

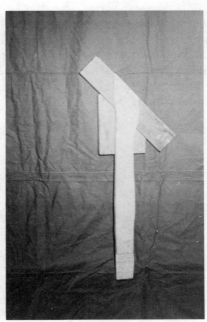

Figure 12-5. Testing Under Vertical Loadings Produced Failure in the Above Joints at 950 Lbs. (Upper Left), 1,030 Lbs. (Upper Right), 1,050 Lbs. (Lower Left) and 1,830 Lbs. (Lower Right)

the University of Kansas and it was confirmed that the joints fail prematurely (Figure 12-5). The tests used an average of four nails per joint, tested in tension. As soon as such a failure takes place, overloading of members will result and subsequent member failures will follow.

More specifically in the case of roof sheathing, it is fastened with six to eight nails per rafter with any of a number of different types. The case of a flat roof produces large suctions at 88.4 mph and would fail in the majority of cases, depending upon the exact type of nail used. The 15° roof produces the largest suctions and sheathing connections would definitely fail if the maximum suction occurs over an entire sheet of plywood. The higher angle roofs are subjected to lower suctions and so sheathing connections should be adequate in these cases.

For a 15° roof house almost all connections would be unable to withstand the forces developed by an 88.4 mph wind. With a 30° roof a fewer number of connections will fail. Even in the case of a 45° roof joints will still fail under the relatively small loading.

It should be noted that the preceeding discussion assumed the wind tunnel loadings. However, if ASCE specification loadings are adopted and the joints checked, many joints will fail, even for the 45° roof.

Summary

It is evident that the typical nailed connections presently in use are woefully inadequate. Even a cursory observation of building techniques reveals that the joints are the weakest portion. Considering the cost of lumber, it is a tremendous waste of resources to build a structure when the material is not able to develop its rated capacity because of poorly constructed connections.

13

Comparative Analysis of Thunderstorms, Tornadoes, Damage Patterns, and Construction Practices

Thunderstorm Structure: Theory and Measurements

Unfortunately, complete measurements of the internal structure of thunderstorms are not currently available. Therefore, it is necessary to combine information from both measurements and theory to provide the most complete picture possible. Thunderstorms are observed to exist in strong wind-shear environments in near vertical position. The only feasible explanation is that the internal structure of the thunderstorm allows its existence in this position. The double vortex thunderstorm model satisfies the internal flow requirements and is consistent with theory and the limited measurements that are available. The first reported measurements from NOAA's Environmental Research Laboratories using dual Doppler radars [49] set up to obtain the flow patterns in thunderstorms are consistent with the presence of a cyclonic vortex in the southern part of the storm and anticyclonic circulation in the northern part (Figure 13-1). This shows airflow in the leading edge of the thunderstorm and is comparable to a horizontal slice through the flow patterns shown in Figure 13-2 for the double vortex thunderstorm. There seems to be little question that a cyclonic vortex exists in the south or southwestern part of severe thunderstorms. The only question concerning the double vortex thunderstorm model is whether the anticyclonic vortex exists in severe thunderstorms. Its magnitude is undoubtedly much less than the cyclonic vortex because of interference of precipitation and opposite circulation to the large-scale cyclonic vorticity, which is generally present with severe thunderstorm activity. Indeed, the precipitation mass may be sufficient to provide the required blockage of environmental winds without a well developed anticyclonic cell. However, anticyclonic circulation within the northeast part of thunderstorms is being reported more and more frequently as Doppler radar techniques are becoming better developed. In addition to the anticyclonic circulation measured by the Wave Propagation Laboratory of NOAA (Figure 13-1) another of the Environmental Research Laboratories, the National Severe Storms Laboratory (NSSL), has measured anticyclonic circulation within several thunderstorms [47,108]. Blockage of the environmental air stream was so complete by some of the severe thunderstorms that anticyclonic vortex shedding has also been observed by NSSL. The relationship between the observed vortex shedding and the double vortex structure is shown in Figure 13-3.

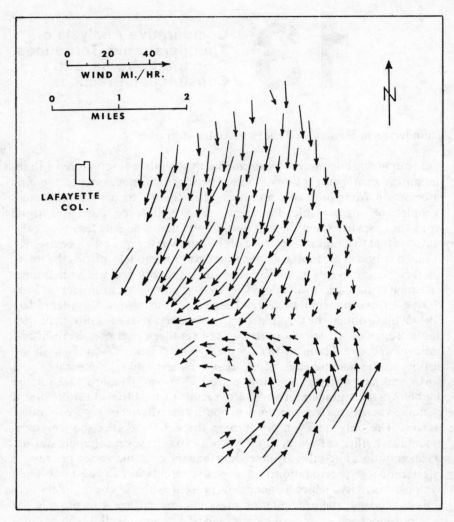

Figure 13-1. National Oceanographic and Atmospheric Administration Dual Doppler Radar Measurements of the Horizontal Wind Field at 200 m Altitude on 29 August 1969 Showing Indications of a Double Vortex Structure in a Colorado Thunderstorm (Ralph Segman [49])

Several other research groups in addition to the Environmental Research Laboratories have been studying horizontal airflow within convective storms by using single or multiple Doppler radar sets. These include the University of Miami and Air Force Cambridge Research Laboratories

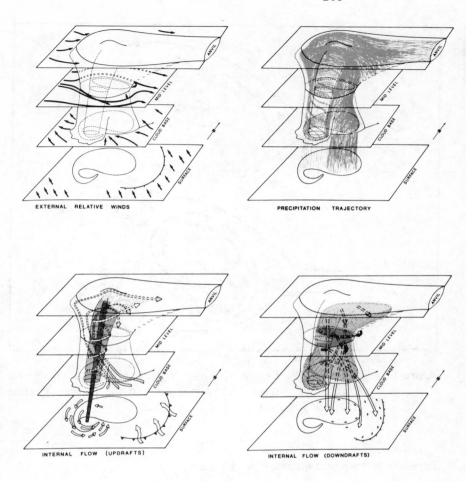

Figure 13-2. The Double Vortex Thunderstorm Model Showing the External Relative Winds That Generate the Precipitation Trajectories, Updrafts and Downdrafts within the Thunderstorm

(AFCRL). Measurements reported by AFCRL [109] also reveal strong evidence for anticyclonic circulation along with cyclonic circulation inside a severe thunderstorm near Boston, Massachusetts.

It is perhaps noteworthy that the double vortex thunderstorm model was developed in 1970 [41] from theoretical considerations before any of the Doppler radar measurements were available to support such a model. The increasingly frequent reporting of both cyclonic and anticyclonic rotation within thunderstorms indicates that the double vortex structure may be the most common internal structure of severe thunderstorms.

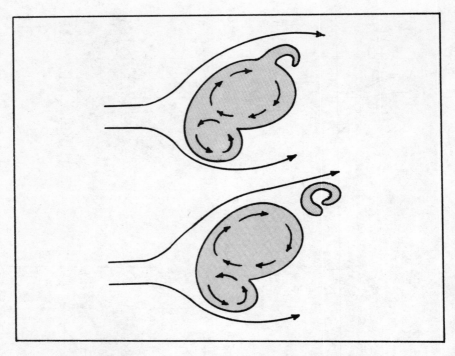

Figure 13-3. Relationship Between Wake Vortex Shedding Observed from Some Severe Thunderstorms and the Double Vortex Interval Structure

Tornadoes: Theory, Simulated, and Natural

Appearance

The tornado vortex theory provides two major regions of airflow: a central core region and a surrounding free vortex region. Whenever particles of dust, water, and other visible materials are present, these two regions are visible. The two regions are shown by photographs to exist in both the unconfined laboratory vortex and natural tornadoes (Figure 13-4). The theory also predicts that the characteristics of the funnel depend upon the strength of both the circulation and the suction in the generation region. The laboratory vortex has confirmed this and has produced a variety of shapes similar to photographs of natural tornadoes. A vortex of large diameter with little upflow was generated in the laboratory when the rotational component was large and the suction was small in the generation

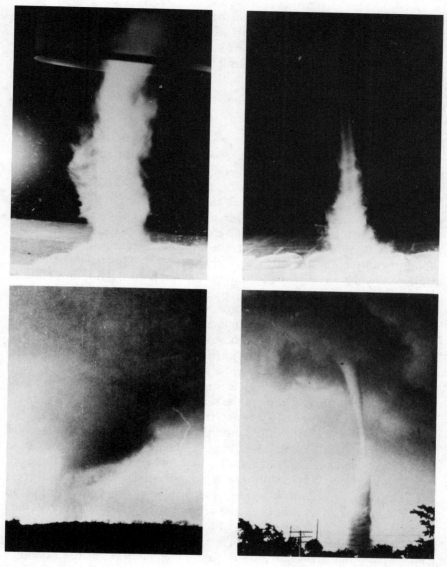

Figure 13-4. Comparison of the Unconfined Laboratory Vortex Produced by Stronger Rotation Than Suction (Upper Left) and Both Rotation and Suction (Upper Right) with Tornado Photographs (The tornado on the left was photographed by Rex Powell a few miles from Lawrence, Kansas, on April 12, 1964. The one on the right was photographed by Leo Ainsworth of National Severe Storms Laboratory near Enid, Oklahoma, on June 5, 1966.)

region. A smaller diameter intense vortex was generated when both the rotational and suctional components were large. With very large suction in the generation region in comparison to the rotation a vortex was sustained. Photographs of natural tornadoes exhibit these same features (Figure 13-4).

Surface Damage Paths

The theory predicts three general types of interaction between the tornado vortex and the surface that will determine its ground track.

The vortex does not stabilize over one location and remain stationary if there is no movement between the vortex and the surface beneath it (Figure 7-10). The laboratory model showed that the base of the vortex wobbles over a sizeable area if there is no relative movement. The vortex describes a series of loops if there is slow relative movement between the vortex and the surface (Figure 7-12). With faster relative movement the vortex travels along the surface in a straight path (Figure 7-13). This behavior can be explained by the air spiraling around the central core and forming a limber rotating cylindrical tube that is in contact with the surface.

As a result of frictional interaction between the vortex and the surface the strong surface core inflow exhibits certain characteristics. First, if the vortex generation region is stationary, the surface inflow is first from one direction and then from the other as the vortex wobbles around over the surface. Second, as the relative velocity increases and the looping path becomes evident, the surface inflow is into the rear of the vortex tube as it loops over the surface. Third, as the relative velocity is further increased and the path becomes nearly straight, the surface inflow is into the rear of the traveling vortex. Both the laboratory vortex and tornadoes exhibit these characteristics. Figure 13-5 shows the damage path of the Topeka and Lubbock tornadoes. The Lubbock tornado, which described a series of loops, was developed by a slow moving thunderstorm. Tornado observers often describe an approaching funnel as swaying back and forth like an elephant's trunk. This would be the apparent motion of an approaching tornado that was forming a series of loops on the surface.

Many tornadoes such as the Topeka tornado (Figure 13-5) produce damage paths that are relatively straight for many miles. Figure 13-6 shows a damage path from one of the April 3, 1974, tornadoes. These tornadoes were developed from fast moving thunderstorms and produced large tornadoes that left damage paths which were visible from the ERTS satellite located over 500 miles in space. The damage path shown in Figure 13-6 from high altitude and ERTS photography extends for many miles across Bankhead National Forest in northern Alabama.

The relationship between different tornado damage paths and a double

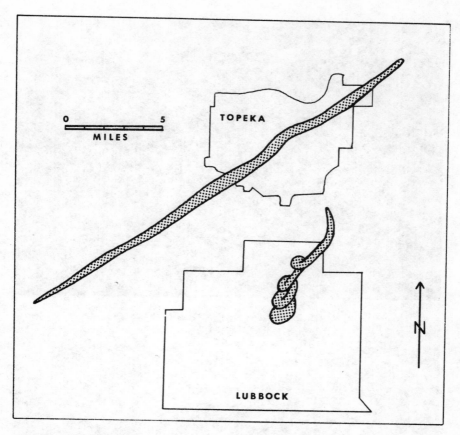

Figure 13-5. Comparison of the Damage Paths Left by the Topeka and Lubbock Tornadoes

vortex thunderstorm is shown in Figure 13-7. A major part of the energy for the tornado must be supplied by the environmental airflow around the supercell thunderstorm. After the mesocyclonic circulation is developed in the southwestern half of the storm the nature of the resulting tornado damage path depends on the speed of movement of the mesocyclone, the strength and location of the tornado funnel within it. If the thunderstorm moves slowly a looping damage path may result from a tornado originating near the center of the mesocyclonic dynamic updraft and a long straight path may result from a faster moving thunderstorm. If the tornado originates on the southern edge of the mesocyclone it will be carried by the mesocyclonic circulation around to the center of the thunderstorm where the environmental winds can no longer feed energy to it and large amounts of precipitation can actively cause its dissipation near the center of the

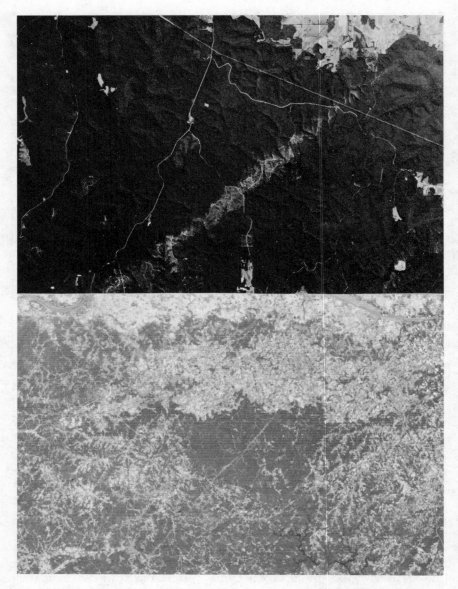

Figure 13-6. A Long Straight Tornado Damage Path Through Bankhead National Forest in Northern Alabama as Revealed by High Altitude (60,000 Ft.) Aircraft Photography (Above) and ERTS Multispectral Scanner, Band 5, Satellite Photography Taken June 3, 1974 (Photograph courtesy of National Aeronautics and Space Administration)

267

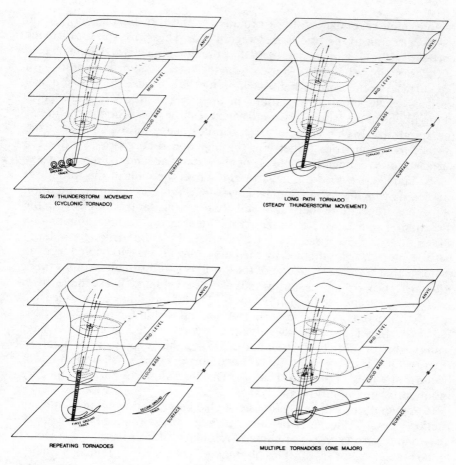

Figure 13-7. Damage Paths and Relationship of the Tornado to the Thunderstorm for Long Path Tornadoes, Looping Tornadoes, Repeating and Multiple Tornadoes

thunderstorm. Multiple tornadoes may originate near the southern edge of the thunderstorm in addition to the central location.

Local Surface Inflow

Observational evidence as well as tornado vortex theory predicts a primary surface airflow pattern resulting from the strong inflow into the tornado vortex core from the rear and rotation of the free vortex region. In addition, in the thin boundary layer region between the core and the free vortex region small secondary votices provide for the transfer of energy from the core region to the free vortex region [72]. These are spiral or Taylor vortices [110,111] and are driven by the core flow, the major damaging part of the tornado. When these spiral or Taylor vortices come in contact with the surface as the core is bent back along the surface, these secondary vortices cause small perturbations in the surface inflow into the core. Photographs of the laboratory model clearly show the secondary nature of the spiral and Taylor vortices. The vortices cause inflow streaks at the surface as the air tries to get into the high speed core flow. Above the surface they are much less intense as they spiral around the main vortex.

A damage pattern from left to right across the path of the tornado is predicted by theory to be slight damage from winds counter to the direction of tornado translation, then rapid buildup to a peak damage intensity in the direction of translation, and then rapidly decreasing damage. Figure 13-8 shows the predicted damage pattern as well as the striking similarity between wax trails produced by the laboratory vortex model and tree fall patterns that resulted from one of the April 3, 1974, tornadoes in northern Alabama. Details of the inflow streaks are evident. These have been previously assumed to be suction spots [22,112] and it has been suggested that their origin is miniature vortices traveling within the tornado vortex. However, they are secondary vortices in the interface region that transfer energy from the core to the free vortex. When this secondary spiral inflow (Figure 13-9) interacts with the surface vortex core, small perturbations occur in the flow of air into the core. These small perturbations and core inflow provide the inflow streaks.

Cycloidial trails in sandy soil have also been used as an indication of suction spots. However, the same tornado that produced the damage shown in Figure 13-8 also produced cycloidal trails in a bare field adjacent to the forest (Figure 13-10). This photograph is enlightening since the major winds were apparently in the direction of movement of the tornado as indicated in the forested area even though cycloidal streaks were left in the soil nearby. Velocity calculations based on the cycloidal marks would, therefore, be too small. Also note that the major damage in the forest is on the right hand side of the cycloidal trails.

Comparison of Model Pressure Mapping, Destruction Model Testing, and Tornado Damage

Pressures obtained from model houses with roofs that were flat, 15°, 30°, and 45° were used to provide external air loadings to predict failures for houses of normal construction. Results for each type of house indicate that the downwind portion of the house can be expected to sustain the least

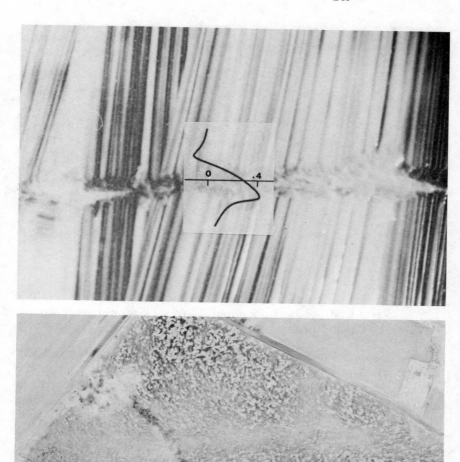

Figure 13-8. Inflow Streaks Associated with the Laboratory Vortex Moving Over a Surface from Left to Right (Upper) and a Tornado Moving Through a Forest in Northern Alabama on April 3, 1974 (Lower photograph courtesy of National Aeronautics and Space Administration)

Figure 13-9. Inflow Streaks and Spiral Vortices (Taylor Vortices) Produced by the Laboratory Vortex Generator with Anticyclonic Rotation

damage (Chapter 10). Wind tunnel destruction tests of model houses with 15° roofs confirmed this prediction (Chapter 11). Systematic evaluation of numerous tornado damaged houses (Chapters 2 and 3) showed that downwind parts of houses were safest most frequently. In addition, the evaluation of tornado damaged houses has shown that within this area the smaller rooms are the most likely to be least damaged. This is to be expected since under normal construction practices the smaller rooms are more strongly constructed per unit area than the large rooms.

The evaluation of tornado damage has shown that varying degrees and types of damage occur to houses within the same area (Figure 13-11). Extensive studies of actual failures occurring as the result of tornadoes revealed similar typical patterns of failure. Apart from the pressure effect as a result of high suctional wind loading, the three most common modes of failure for residential buildings are failure of the wall to foundation connections, failure of the roof to the wall connections, and failure of the roof itself.

It has been observed that in the same area with the same intensity of wind loading, different houses exhibit weaknesses resulting in one of the above mentioned modes of failure.

Figure 13-12 shows a typical case of wall to foundation failure. Apparently the superstructure as a whole has been pulled or sheared off the foundation due to insufficient strength of the connections.

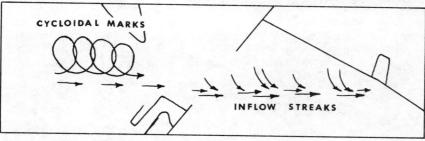

Figure 13-10. Cycloidal Marks in a Bare Field and Inflow Streaks in a Forest in Northern Alabama, April 3, 1974 (Photograph courtesy of National Aeronautics and Space Administration)

Figure 13-13 shows a typical case of roof to wall failure. The roof as a whole has been torn loose from the wall including the bottom chords. The actual cause of the failure depends on construction practices. Anchorage of the roof trusses to the wall usually is provided by toe-nailing the bottom chords to the wall plate. This type of construction is weak against uplifting forces caused by suction acting on the roof.

Figure 13-14 shows a typical case of roof failure. The bottom chords are still connected to the walls, but the remainder of the roof has been torn off. It is often difficult from observed damage to determine the actual sequence of failure. Figure 13-14 shows the onset of roof damage to be failure of the roofing material and decking.

Figure 13-11. Comparison of Tornado Damaged Houses (Upper) and a Model House Partially Destroyed in the Wind Tunnel. Damaging Winds were from Left to Right. (Photographs by J. R. Eagleman)

Figure 13-12. Example of a Wall to Foundation Failure

Figure 13-13. Example of a Roof to Wall Failure

Figure 13-14. Example of Roof Failure

A typical type of failure consisted of failure of roofing on the leeward side, followed by tearing loose of the roof decking from the top chords, leaving the trusses more or less intact.

Another failure mode consisted of failure of roofing and decking, first, on the leeward side, followed by roof deck failure on the windward side, followed by whole sections of the roof decking pulling parts of the truss members with them—either by breaking them or causing failure of the joints. This type of failure often occurred as the roof folded open by hinging about the leeward wall support. This has particularly been observed when the roof has sizeable eaves. The uplifting pressure acting along the bottom of the windward eaves combined with a suction acting along the top particularly in the case of a weak roof to wall connection causes the roof, as a whole, to fold over, thereby taking the bottom chord with it or partially or fully breaking away from it.

The wind tunnel pressure measurements give an explanation of many of the anomalies. The pressure study showed that the orientation of the house relative to the wind, the roof angle, and appendages such as dormers, porches, and chimneys were important factors in determining the air loadings on the houses. In general the steeper roofs were lighter loaded; the more appendages, the lower the local air loading.

The destruction tests provided an explanation for some of the anomolies in damage due to various types of venting of houses. If the model house had openings on the windward walls of the house, it was highly susceptible to damage. If the downwind walls only were vented, it was less susceptible to damage. However, a completely air tight model house withstood higher wind loadings than either of these two. The most successful venting was provided when the area on the roof just downwind of the peak was vented. When this was combined with venting in the ceiling, a marked increase in the ability of the model house to withstand wind loadings was observed.

Figure 13-15 shows a comparison between tornado damaged houses and model house damage. The center house, which was damaged by the Lubbock tornado, was typical of several observed both in the field and laboratory. In this house the safest location on the first floor would have been near the inside walls of rooms opposite the approach direction of the tornado.

The pressure loadings and destruction tests also predict the different types of failure of houses based upon normal construction practices. For example, the investigation of a number of newly constructed homes showed the same types of weakness: wall to foundation connections, wall to roof connections, and roof truss connections.

The wind tunnel results applied to houses with these structural weaknesses indicate that for houses with low roof angles, the following types of damage occur: local roof damage at upwind edge or behind the peak, roof peeled off and upwind walls caved in, or house moved downwind from the foundation. Houses with high roofs might be expected to exhibit the following types of damage: local damage downwind of the peak, roof slid off the walls, house slid off the foundation, or upwind portion caved in. Observations of tornado damaged houses have shown all these types of damage. Figure 13-16 shows a common type of damage occurring to the peak of a roof during a tornado as well as a model roof damaged in the wind tunnel. These show similar wind effects as predicted by the pressure distribution.

Building Criteria

Damaging Winds

Observations of tornado damage, destruction model testing, model pressure mapping, tornado inflow streaks, laboratory model tornado tracks, and theory all demonstrate that the damaging winds in the tornado are in the direction of tornado movement. The principal component of this wind is a result of the inflow to the core from the rear. Since accurate velocity measurements have never been made in tornadoes and the intensity of

Figure 13-15. Houses Damaged by the Lubbock Tornado (Upper) and a Model House Partially Destroyed in the Wind Tunnel. Damaging winds were from left to right in both cases.

Figure 13-16. Comparison of Wind Damage to Houses by a Tornado and in the Laboratory with Roof Failure at the Peak as Predicted by the Pressure Distribution Measured in the Laboratory

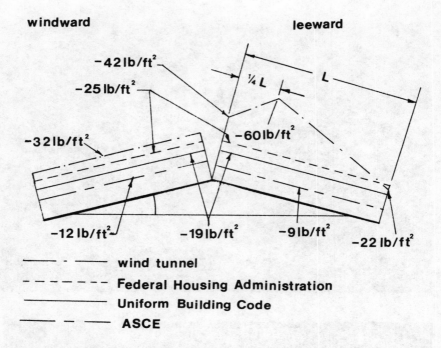

Figure 13-17. Comparison of Various Building Code Specifications with Pressure Measurements Over Model Houses in the Laboratory Converted to 88.4 mph

storms vary, estimates of wind speeds vary greatly. From observations made by the authors, the maximum core inflow velocity in large tornadoes is 300 to 600 mph.

For any selected wind speed the pressure loading on a specific building may be calculated from pressure coefficients obtained from wind tunnel models (Chapter 10). If it is desired to have a structure withstand all but the central core flow winds of a moderate to large tornado, it would appear that the loadings should be calculated for approximately a 300 mph wind speed. To withstand the very large and severe storms the wind loads should probably be based upon winds of the order of 600 mph. However, a design velocity of 200 mph would allow buildings to withstand the greatest winds in most tornadoes.

Building Codes

It is evident from the discussion in Chapter 12 that current building practice is inadequate to withstand the true forces generated in an 88 mph design

Figure 13-18. House That was Lifted off Its Foundation and Dropped in the Back Yard During the Topeka Tornado (Photograph by J. R. Eagleman)

wind velocity. While part of the inadequacy is caused by poor design details and construction practices, part of the problem results from inadequate code provisions. The codes [113, 114] do not accurately project wind in-duced forces and also fail to specify types of construction. In addition, they are often not followed or enforced. The biggest problem with the building codes is that the specified pressure and suction values are not accurate as has been demonstrated by the wind tunnel testing. For comparison the pressure values for a 15° roof as set forth by various specifying agencies are shown in Figure 13-17. It is obvious that, in general, building codes do not accurately portray the wind forces on homes. It should be noted that the discrepancies are worst for the low-angle roofed structures and that there is less discrepancy for the steeper (45°) angle roofs.

Construction Practices

It appears that many of the home-building methods presently used rely upon the dead weight of the structure to join members. When such a structure is subjected to uplifting forces such as occur in strong winds, failure will take place. Figure 13-18 is an example of a poured concrete

Figure 13-19. Poor Truss Design Because of Unclinched Nails

Figure 13-20. Poor Truss Design Because of Notched Members

foundation that was apparently not bolted to the wall plates with the assumption that the weight of the house would hold it in place. As stated previously in Chapter 12, many joints fail even under the relatively low loads specified by common building specifications. It is obvious that current construction practice ignores design for wind. The Uniform Building Code in section 2307(c) states that: "Adequate anchorage of the roof to walls and columns, and of walls and columns to the foundations to resist overturning, uplift, and sliding, shall be provided in all cases" [113]. As can be seen from the prior discussion, this section of the codes is hardly satisfied. It is evident that apart from the fact that the codes are not completely satisfactory, current construction practices ignore existing code provisions.

The inadequacy of many of the current construction practices becomes clear upon inspection of houses under construction. Nails are often driven into the end grain of members and, therefore, really cannot be counted upon to carry any load at all. Usually there are only enough nails at any joint to just hold the members in place during erection. As can be seen in Figure 13-19 the nails protrude and are not clinched, while in Figure 13-20 members are notched, thus reducing the bearing areas and inefficiently utilizing the material. Besides nailed joints, the foundation bolts that are to resist uplift forces are very often not to be found, since they were not attached during construction (Figure 13-18). From the above discussion it becomes clear why the damage assumes the three distinct patterns discussed previously.

14 Practical Applications

Tornado Preparedness

Tornado Watches and Warnings

The most common psychological defense from the tornado consists of ignoring it and assuming that it will only affect someone else. While the probability of a tornado striking a given spot is only one in a few hundred years, even in the central United States, this is of little help if a large funnel is approaching. The chances for surviving a major tornado can be substantially improved by preparation well in advance as well as during a tornado warning. Advance preparation consists of constructing a tornado shelter, obtaining knowledge of safest locations in houses, and attention to building design and construction if this is to be the only shelter from severe storms.

Tornado watches are issued by the National Severe Storms Forecasting Center (NSSFC) in Kansas City. It is currently impossible to predict exactly where or when an individual thunderstorm will occur [115]. But it is possible to predict general areas where severe thunderstorm development is likely based on atmospheric conditions. The average dimension of a tornado watch is 140 by 175 miles, covering about 25,000 square miles. The valid time is generally six hours and it is the goal of NSSFC to issue the watch at least one hour in advance of tornado development. Over the past 21 years only 27 percent of the tornadoes have occurred in watch areas [115]. However, the percentage of large severe tornadoes that are forecasted is much better than for small weaker tornadoes. About half the total number of tornadoes that occur each year are very short lived, may be only 100 feet in width, and stay on the ground for a mile or less. These tornadoes are difficult to forecast since the general weather patterns do not indicate severe weather activity and most of them do not show up on radar, since the hook echo is not present. They may be only vortices caused by wind shear on the edge of thunderstorms that do not have a double vortex internal structure and, as such, may last for less than five minutes.

The average and very large tornadoes are normally associated with general weather patterns allowing them to be forecasted much easier. They also are more likely to be identified because of a hook echo on the radar screen. Fifty-six percent of tornadoes that produced fatalities were fore-

283

casted and 66 percent of all deaths from tornadoes between 1952 and 1972 occurred in tornado watch areas.

Tornado warnings are issued only after a tornado has been sighted or indicated on radar and are the responsibility of local National Weather Service offices. Adequacy of the tornado warning system varies from community to community. Many counties have volunteer storm spotter organizations. Civil defense groups organized under the Civil Defense Preparedness Agency are an important part of the warning system in larger urban areas. Rapid communication is essential for a successful tornado warning system with local radio and television stations disseminating information to the public. Some cities have a network of sirens to further alert the community to severe weather activity.

As the effectiveness of tornado forecasts and warning systems increases additional time is provided for seeking shelter. During a tornado in Salina, Kansas, in October 1973, a trailer court was completely demolished (Figure 14-1) but there was not a single death because 80 people had gathered in an underground shelter.

Underground Shelters

For many years the tornado shelter was an integral part of many prairie family lives. This shelter was usually separate from the house and constructed with sturdy walls and roof and usually had a north entrance. It also served the additional function of food storage. However, the tornado shelter is disappearing from the central United States where tornadoes are most common, in spite of some efforts to provide modern designs for outdoor storm shelters [116].

Underground shelters with an inside entrance are preferable because of the short warning time for most tornadoes. Such shelters can be easily constructed at the time the house is being built by simply pouring a little additional reinforced concrete. If it is a part of the basement it should consist of concrete walls eight inches thick with a top roof slab of eight inches. These should be continuous and reinforced with metal bars at nine-inch intervals. As an extra precaution the entrance should consist of a concrete hallway with a 90° turn or some other arrangement that would prevent flying debris from reaching the interior of the shelter.

Tornado Safety

Most people are not sufficiently concerned about tornadoes to construct storm shelters. They are more concerned about which part of their house is

285

Figure 14-1. Tornado Damage to a Mobile Home Park in Salina, Kansas, Where All the Occupants Were Unharmed Because They Were in a Common Shelter Provided by the Park (Photograph courtesy of Evelyn Burger, Salina Journal)

safest in case of a tornado. Before the investigations described in previous chapters were conducted, people were advised to seek shelter in the southwest corner of the basement. It was also assumed by many people who lived in houses without basements that the southwest part was also the safest. This has been shown statistically to be the worst location in houses without basements. Four of the six storms investigated had the highest percentage of safe locations in the part of the house opposite the approach direction of the tornado. The other two were safest in the central part of the houses. It was also found that room size was important, with the ranking from most safe to unsafe consisting of closets, bathrooms, hallways, small rooms, and large rooms. The investigations of tornado damage indicate safest locations in the following priority:

1. Inside underground shelter.
2. Outdoor underground shelter.
3. Poured concrete basement with no windows. Available data indicate that safest locations are opposite the approach direction of the tornado. Since most tornadoes come from the southwest the northeast part of basements is recommended because of the statistical evidence as well as observations that any remaining ceiling after severe damage is more likely to be located over this section.
4. Poured concrete basement with small windows opening to the outside. The recommended location for shelter is the northeast part for a tornado approaching from the southwest because this is the area with the greatest probability of floor remaining overhead and the area with the smallest probability of debris coming through the windows.
5. Half basement. Houses on a slope with only half basements are safest in the most submerged part of the basement.
6. Stone, brick, or block basements. The northeast part of these basements were safer more frequently than any part along the south because of the tendency for movement of the whole house in the direction of movement of the funnel, which was from the southwest for five of the six major storms investigated. When the house moved toward the northeast, debris composed of boards, pipes, and stones accumulated in the south part of the basement.
7. Above ground rooms. Frequently, the safest locations were closets and small rooms in the northeast part of the house on the first floor. Rooms facing the approach direction of the tornado, which is usually from the southwest, were the most unsafe. The northeast part of the first floor of a two-story house with a strong foundation-wall connection should be safer than the first floor of a single-story house. Small rooms in the central part of the house were frequently safe and may be the best choice if the approach direction of the funnel is uncertain.

8. Automobiles. Automobiles are safe only if used to get out of the path of the funnel, which usually travels at about 30 mph. Otherwise, they may offer very little protection since they can be lifted by the winds and smashed with great force against other objects.

9. Mobile homes. Mobile homes usually offered no protection from tornadoes. A ditch located away from electric lines would normally be safer.

10. Open country. Try to find a small depression or ditch running perpendicular to the direction the tornado is moving. Lying flat on the ground and holding on to anything available may also help. A bridge or culvert may be used for protection from a tornado.

11. In addition, the following should be considered as appropriate with the above:

a) Flying objects. All types of boards, pipes, cans, and other loose objects can readily become missiles during a tornado (Figure 14-2).

b) Windows and doors. Model tests indicate the house may be expected to withstand the wind pressures best with doors and windows all closed. Upwind venting makes a house particularly vulnerable to damage. Although downwind venting of the model houses also caused them to be damaged sooner than unvented models, it is not certain that all the pressure changes, such as may be caused by venting or leakage during a tornado, were accurately modeled. It was demonstrated, however, that proper roof venting reduced damage from high winds.

Prior Planning Through Better Design and Construction of Houses

General

Home-building often has been more a matter of expedience and styling than the result of a conscious design process. The practice of using expensive lumber in homes, and connecting it poorly, clearly illustrates the problem. The lumber, in houses as they are now built, can never attain its full capacity. Thus, a more balanced design and more efficient use of materials would result from improvement of the presently inadequate joints.

While previous discussion has centered on approximately a 90 mph wind velocity design, it would be feasible to design for up to 200 mph winds if no factor of safety is used. It most certainly will be possible to expect a factor of safety, on the average, of about 2.5 in the improved joint design,

Figure 14-2. Small Objects Become Missiles During a Tornado as Shown by the Penetration of This Silver Spoon During a Tornado in Tulsa, Oklahoma, in 1973 (Photograph courtesy of Lewis Jarrett, Tulsa Tribune)

using common building materials, for a wind velocity of 90 mph. It is only the inherent inefficiency of present unbalanced design practices that prevent such drastic improvements in the structural performance of houses.

Building Codes

As seen from the comparisons shown in Figure 13-17, wind forces as predicted by most building codes are inadequate. In most cases the codes underpredict the loads and for a 15° roof angle they not only drastically underpredict the loads, but also predict other loads that do not exist. A closer approximation of actual wind suctions or pressures for the 90 mph design velocity in the cases studied would be:

1. Flat roof; 35 PSF suction over entire surface
2. 15° roof; on leeward slope a suction of 50 PSF at the peak varying linearly to 20 PSF at eave, and a 30 PSF suction on the windward slope
3. 30° roof; 10 PSF pressure to a 20 PSF suction on windward slope, and a suction of 25 PSF on leeward slope
4. 45° roof; a 15 PSF pressure to a 10 PSF suction on windward slope, and a 20 PSF suction on the leeward slope
5. A 20 PSF pressure on windward wall and a 20 PSF suction on leeward wall

It can be seen that wind loads specified by codes have been too low for the low angle roofs. Therefore, an improvement in the building codes is definitely needed, especially for low roof angles.

The Uniform Building Code (1973) requires construction to be such that uplift forces will be resisted. This leaves much to be desired since nothing specific is said about type of connections, the number of connectors, or any other details. Ambiguities such as these have led to the most expedient and probably least satisfactory types of construction.

Architectural Design of Houses

The architectural design of a house determines to a large extent the wind loadings that will be placed upon it during a tornado. These studies indicate the following items should be considered:

1. Orientation. For each particular type of house there is one orientation relative to the wind that yields the least wind loading. The house should be orientated at this angle for a southwest wind, if practical. For rectangular houses with flat or 15° roofs the short side should be turned toward

the southwest. If the roof angle is 30° the long side should be turned toward the southwest and for a house with a 45° roof the angle is not as critical, but the optimum orientation is with the short side turned toward the south or slightly toward the southwest.

2. Slope of the roof. Wind loadings increase as the roof angle decreases. The 45° roof provides the least loadings of those investigated. Therefore, steeper roofs should be considered for added resistance to high winds.

3. Dormers, porches, etc. The use of dormers, porches, and other irregularities in roof line appears to reduce the peak loadings.

4. Venting. It is highly advisable to vent the inside of the house from the downwind side of the roof peak. This allows a large negative pressure inside the structure. The attic should also be vented through the ceiling to the lower part of the house to facilitate the negative pressure equilization through the house. Negative pressure inside effectively reduces the loading on the roof, rear, and at least two sides of the house. Although at the same time an increased loading occurs on the upwind two sides of the house; the critical loadings are reduced.

Structural Framing Members

For a 90 mph design wind velocity with a factor of safety on the average of 2.5, a No. 2 southern pine or douglas fir should be entirely adequate. This is assuming that the typical lumber sizes such as 2 × 4 and 2 × 6 are used. For roof trusses of the higher angles bracing requirements of the internal members should be checked to assure that the members will develop the expected forces.

It should be entirely feasible that if no factor of safety is assumed, the members in typical designs could be expected to perform adequately, or at least suffer minimum damage under wind velocities of up to 200 mph. Beyond this velocity typical construction practices, member spacing and member sizes become inadequate. Thus, typical wood construction would not suffice. For higher wind velocities alternative methods and materials would have to be used.

For the 200 mph wind the present type of construction, without a factor of safety, is inadequate. In general it is safe to say that for the low angle roofs the top and bottom chords of the roof trusses would have to be of a much higher grade of lumber (dense structural) to withstand 200 mph wind forces while still using a 2 × 6 member. It would be best to use a dense structural grade of lumber in these top and bottom chords of the trusses for all roof angles. Also, a structural grade of plywood should be used for the roof sheathing to assure that bending stresses do not reach ultimate, if a two-foot center to center truss spacing is maintained. The best method would be an analysis of the particular structure and the use of high capacity members where needed and lower grade members elsewhere, so costs may be kept reasonable.

Connections

It is at the connections that present construction methods need the most improvement. For the 90 mph design wind and an average factor of safety of 2.5 almost all structural joints for all roof angle configurations need improvement.

For attaching the plywood roof sheathing a better nail spacing would be three inches to four inches, assuming there is little standardization of nail types. This nail spacing could be increased if deformed shank or coated nails with higher withdrawal resistances were used. For truss member connections the best choice would be split ring connectors. Nailed gusset-plates could be used but the number of nails and size of the joint soon becomes prohibitive. These same problems would be true of bolted, gusset-plate joints, although they may be more satisfactory than nails. Besides structural considerations, split ring connectors would require less fabrication time than gusseted joints.

For truss to wall connections a combination split ring, shear plate, and bolted angle (Figure 14-3) arrangement would be good. As can be seen from tornado damage patterns this connection is particularly important. For wall stud connections to baseplates, rather than toe-nailing or nailing on end grain, a much better connection would be metal angles. Some of these are available commercially. As was mentioned in Chapter 13 most houses are weakly connected to their foundation and may only be held down by their own weight. It is evident from observed damage patterns that the wall to foundation is a particularly vulnerable connection. To improve this connection a one foot spacing between bolts would be more realistic. However, some improvement would result from a more conscientious attachment of washers and nuts.

Concluding Remarks

These suggestions should be considered only as a guide to the type of improvements needed. If the recommendations on connections are followed it should be possible for houses to withstand winds of up to 200 mph,

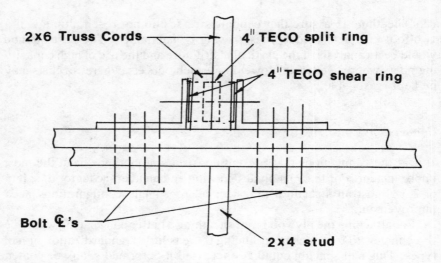

Figure 14-3. Suggested Improvement in Truss to Wall Connection

assuming no factor of safety and short duration loading. Thus, the benefits gained from such improvements will result in a tremendous increase in the overall structural capacity and safety of a typical home.

If the public is properly informed concerning the safest locations in houses there should be fewer fatalities and injuries during tornadoes. It is particularly important that people living in houses without basements (in the southern United States, for example, where houses seldom have basements) do not seek shelter in the southwestern part of the first floor. A survey of several hundred people in Lawrence, Kansas, in 1974 indicated that almost two-thirds of them thought the southwest part of the first floor was the safest location during tornadoes.

Appendix A:
List of Frequently Used
Symbols

Symbols

A Vortex core area

C Coefficient

D Drag Force

E Efficiency factor

F Force

I Intensity factor

K Roughness Height

L Lift force

M Mach number

$\mathcal{M}$ Mass flow

R Gas constant

R Reynold's number

Q Characteristic area

S Speed index or side force

T Temperature

U Uriel number (blockage factor)

WA Wind angle

a Speed of sound

b Vortex core length

g Gravitational constant

h Height

k Constant

p Pressure

q Dynamic pressure

293

r Radius

s Path length

u,v,w Components of velocity in x, y, z directions

v Velocity

Γ Circulation ($2\pi v_\theta$)

ω Angular velocity

ρ Air density

γ Ratio of specific heats (Air—1.4)

μ Absolute viscosity

ℓ Mixing length

ν Kinematic viscosity

δ Boundary layer thickness

Subscripts

A Atmospheric conditions

B Blockage

D Drag

F Force

GL Ground level

L Lift

T Translation

o Center of vortex core or incompressible

c Inner radius of free vortex flow

e Effective

i Induced

j Jet stream

m Mesocyclonic vortex

p pressure

r Radial component

t Total

w *z* component

θ Tangential component

Γ Circulation

∞ Free stream conditions

Exponents

α

β

δ

ε

ζ

λ

m

n

Glossary

Air mass—A term applied by meteorologists to an extensive body of air within which conditions of moisture and temperature in a horizontal plane are essentially uniform.

Analog computer—Also analog field plotter; the flow of electricity through conductive paper can be used to model other types of flow patterns.

Backing winds—A wind profile where the wind direction changes in a counterclockwise direction with increasing height.

Boundary layer—A small region next to the surface of a body in a moving fluid in which the fluid flow speed is reduced by the viscosity of the fluid.

Chord—Top or bottom main members of the truss framework.

Circulation—The product of the velocity component along the flow path and the length along the flow path.

Clinching—Practice of bending over protruding nail ends by hammering.

Coefficient—A non dimensional number which expresses certain characteristics of a body.

Cold front—The line of discontinuity at the forward or leading edge of an advancing cold air mass which is displacing warmer air.

Collar cloud—A circular visible part of the base of a thunderstorm corresponding to the rotating mesocyclonic vortex within a severe thunderstorm. The tornado funnel is smaller and located within the collar cloud.

Compressible flow—Air flowing at a velocity high enough that the changes in air density cannot be ignored without introducing a large error in the results.

Convection—The upward and downward movement, usually thermally induced, of a limited portion of the atmosphere. Convection is an essential part of cumulus cloud formation.

Core flow—The air flowing along the vortex axis in the region of solid body rotation in the central part of the vortex.

Cumulonimbus—A type of cloud which is developing vertically (thunderstorm).

Cyclone—Any atmospheric vortex rotating in a counterclockwise direction in the Northern Hemisphere, clockwise in the southern. In different localities it can mean a tornado, hurricane, or midlatitude cyclone.

Cyclonic—Counterclockwise rotation of vortices in the Northern Hemisphere.

Dead load—Permanent loading mainly consisting of the weight of the structure itself.

Density—The mass of a substance per unit volume.

Dew Point—The temperature at which condensation begins in a cooling mass of air. Dew point varies with specific humidity.

Disturbance—A local departure from the normal or average wind conditions in any part of the world. Disturbance may be synonymous with cyclone, or areas of barometric depression.

Divergence—In the atmosphere, if streamlines describing airflow spread apart as viewed from above or below, horizontal divergence is indicated. Velocity divergence exists in westerly airflow if the wind velocity to the east of a given location is greater than the wind velocity to the west.

Drag—The force parallel to the relative wind which resists the motion of a body moving through the air.

Dry adiabatic lapse rate—The calculated rate of cooling as air is lifted in the atmosphere (10°C/km). This is also the rate of heating due to higher pressure as air descends in the atmosphere.

Dry line—The western boundary of the moisture tongue which separates the warm dry air to the west from the warm humid air to the east. This line is frequently related to thunderstorm development.

Dust devil—A whirlwind which develops from surface heating and is common over hot dry land. It is typically a small weak atmospheric vortex.

Dynamic pressure—The kinetic energy of a unit volume of air.

Dynamic updraft—The updraft core inside the mesocyclonic vortex which is an extension of the tornado updraft core.

Eaves—Overhanging section of the roof.

Echo free vault—The regions within a severe thunderstorm where no return signal is received by radar because of absence of precipitation.

ERL—Environmental Research Laboratories of NOAA which includes the National Severe Storms Laboratory, the Wave Propagation Laboratory in Boulder, Colarado, the National Hurricane Center in Miami, Florida, as well as other NOAA laboratories.

Factor of safety—Ratio of critical stress to working stress.

Free stream—The conditions existing upstream in the undisturbed air flow.

Gas Constant—The proportionality factor relating air pressure, density, and temperature.

Gusset plate—Plate connecting individual truss members.

Gust front—The cooler air associated with the downdraft of a thunderstorm which spreads out as it reaches the surface much like a miniature cold front.

Hook echo—The unique shaped radar return from severe thunderstorms.

Hurricane—A tropical storm larger in size than a tornado but less intense. Hurricanes are related to warm ocean temperatures but extend through the troposphere. Winds are greater than 74 mph.

Incompressible flow—Air flowing at sufficiently low velocity that the air density may be considered constant.

Inflow streaks—Streaks of more severe damage from tornadoes because of concentrated inflow into the core of the tornado.

Intensity factor—A nondimensional number that expresses the combined strength of the flow parameters that form and sustain a tornado vortex.

Inversion—The temperature of the air usually becomes lower with increasing height. Occasionally, however, this normal condition is reversed. When the temperature increases with height there is said to be an inversion.

Jet stream—A stream of air flowing from west to east around the earth at about 40,000 ft. with an average velocity of 40 mph in summer and 80 mph in winter.

Kutta-Joukowski force—A force arising from rotation such that cyclonic rotation within a thunderstorm causes curving to the right. Same as the magnus force.

Lapse rate—The amount of temperature decrease with height. A negative lapse rate or inversion exists if the temperature increases with height.

Leeward—The side of a house or object opposite the direction from which the wind is coming is the leeward side.

Lift—The force perpendicular to the relative wind produced by a body.

Lifted index—An index used to specify the stability of the atmosphere with positive index values stable and negative values unstable.

Low level jet—A band of air having a slightly higher velocity than in surrounding regions. It normally flows northward around a midlatitude cyclone and may accompany a moisture tongue from the south.

Mach number—The ratio of the local speed to the speed of sound.

Magnus force—A force originating from rotation which causes deflection. A cyclonic rotating thunderstorm curves to the right and one with anticyclonic rotation curves to the left.

Mesocyclone—Within a severe thunderstorm the southwestern half of the thunderstorm with cyclonic rotation is the mesocyclone. The tornado is a smaller vortex within the mesocyclone.

Midlatitude cyclone—Large centers of low pressure with counterclockwise rotating air. These large storms normally have a cold as well as a warm front and come from the west in the Northern Hemisphere.

Moisture tongue—A stream of warm moist air from the Gulf of Mexico flowing northward around a low pressure center.

NASA—National Aeronautics and Space Administration.

NOAA—National Oceanographic and Atmospheric Administration.

Non-viscous—A region of flow in which the effects of viscosity are small and may be neglected.

NSSFC—The National Severe Storms Forecasting Center located in Kansas City, Missouri.

NSSL—National Severe Storms Laboratory located in Norman, Oklahoma, one of NOAA's laboratories.

Pressure—force per unit area.

Rafter—Roof truss main member.

Rawinsonde—Technique used by the National Weather Service to measure the existing winds at various heights throughout the atmosphere.

Relative wind—The wind direction and velocity relative to a moving thunderstorm.

Reynold's number—A non-dimensional number which expresses the ratio of the inertia forces to the viscous forces in an airflow.

Roof sheathing or decking—Usually 4' x 8' plywood sheets nailed to top chords of roof trusses for support of roofing material.

Roughness height—Height above the surface of various protrusions which will affect the physical properties of airflow.

Shear index—An index related to severe thunderstorm development obtained by determining the number of layers in the atmosphere where the low level relative winds are opposed by the mid to upper level relative winds.

Shear ring split connectors—Metal rings and bolts for connecting wooden members.

Side force—The force perpendicular to the wind and the lift force.

Simply supported—Free to rotate at both ends and free to move horizontally at one end only.

Sink—In the atmosphere a sink exists if there is a deficiency of air.

Source—In the atmosphere a source exists where there is surplus air.

Speed index—A non-dimensional number which expresses the intensity of a vortex at any point along its axis.

Squall line—Thunderstorms that occur in a line which may contribute to their intensification.

Static pressure—The local or ambient pressure in an air flow.

Synoptic patterns—Distribution of weather variables over a large area for a given time.

Thermodynamics—Action caused by heating.

Thermal updraft—The warm moist air flowing into thunderstorms at low levels that provides the double vortex structure inside severe thunderstorms.

Tornado—The most intense vortex in the atmosphere having a central updraft core and cyclonic rotation in the Northern Hemisphere, although a few rotate anti-cyclonically.

Tornado Cyclone—Another name for the mesocyclonic vortex within the thunderstorm.

Trough—As the air within the troposphere flows around the earth there are regions where the air stream bends in a counterclockwise direction that is called a trough.

Truss—Combination of beams or boards to form triangles which support the roof of houses.

Unstable atmosphere—Increased convection and vertical motion in the atmosphere because of surface heating and decreasing temperature with height.

Upper air inversion—An inversion (region where the temperature increases with height) frequently exists at about the 850 mb level when severe thunderstorms develop.

Upwind—The side of a house or obstacle facing the direction from which the wind is coming is the upwind side.

Veering winds—A wind profile where the wind direction changes in a clockwise direction with increasing height.

Velocity—The rate of change of a substance or object through space with time in a specific direction.

Viscosity—The internal flow resistance of a fluid.

Viscous—The region in a fluid flow in which the effects of viscosity should not be ignored.

Vortex—Spiraling air around a central core. Midlatitude cyclones, hurricanes, tornadoes, waterspouts, and dust devils are vortices of different size and intensity.

Vortex, bound—A vortex that is confined as in the generation region.

Vortex filament—A line representing the vortex core.

Vortex, free—A vortex that is unconfined.

Vortex, trailing—A vortex that trails from a bound vortex.

Wall studs—Vertical wall members usually 2″ x 4″ spaced at 16 inches.

Warm front—The discontinuity at the forward edge of an advancing current of relatively warm air which is displacing a retreating colder air mass.

Waterspout—A vortex thought to be similar to a small tornado except that it occurs over water and generally from a smaller cloud.

Wind angle—Wind direction expressed as an angle from 0 to 360° with 0° being north, 90° being east, etc. Also, the angle between the wind approaching a body and a reference line on the body.

Wind shear—Winds in adjacent vertical layers in the atmosphere may occur from different directions resulting in wind shear between the layers.

Windward side—Side that the wind is flowing against.

Bibliography

[1] Segner, E.P. Jr., "Estimates of Minimum Wind Forces Causing Structural Damage," U.S. Weather Bureau research paper no. 41, 1960, pp. 169-75.

[2] Flora, S.D., *Tornadoes of the United States,* University of Oklahoma Press, Norman, Okla., 1953, 194 pp.

[3] Fujita, T.T., "Tornadoes Around the World," *Weatherwise,* 1973, vol. 26, pp. 56-62, 79-83.

[4] Grey, W.M., "Research Methodology, Observation and Ideas on Tornado Genesis," *Preprints, Seventh Conference on Severe Local Storms,* American Meteorological Society, Boston, Mass., 1971, pp. 292-98.

[5] Battan, L.J., *The Nature of Violent Storms,* Doubleday & Co., Inc., Garden City, N.Y., 1961, 158 pp.

[6] Hall, R.S., "Inside a Texas Tornado," *Weatherwise,* 1951, vol. 4, no. 3, pp. 54-57, 65.

[7] Fujita, T.T., D.L. Bradbury, and C.F. Van Thullenar, "Palm Sunday Tornadoes of April 11, 1965," *Monthly Weather Review,* 1970, vol. 98, no. 1, pp. 29-69.

[8] Barnes, S.L., "Morphology of Two Tornadic Storms: An Analysis of NSSL Data on April 30, 1970," *Papers on Oklahoma Thunderstorms, April 29-30, 1970,* NOAA Technical Memorandum ERL NSSL-69, 1974, pp. 125-39.

[9] Brown, R.A., D.W. Burgess, and K.C. Crawford, "Twin Tornado Cyclones within a Severe Thunderstorm: Single Dopplar Radar Observations," *Weatherwise,* 1973, vol. 26, no. 2, pp. 63-69, 71.

[10] Donaldson, R.J., "Vortex Signature Recognition by a Doppler Radar," *Journal of Applied Meteorology,* 1970, vol. 9, no. 4, pp. 661-70.

[11] Canipe, Y.J., J.E. Vogel, and R.A. Clark, "Detailed Analysis of Tornado-Producing Thunderstorms Using Digital Radar," *Preprints, Eighth Conference on Severe Local Storms,* American Meteorological Society, Boston, Mass., 1973, pp. 57-60.

[12] Biggs, W.G. and P.J. Waite, "Can TV Really Detect Tornadoes?" *Weatherwise,* 1970, vol. 23, pp. 120-25.

[13] Bradshaw, H. and V. Bradshaw, "How You Can 'See' Tornadoes on TV," *Popular Mechanics,* March 1969, vol. 131, pp. 93-96.

[14] Stranford, J.L., M.A. Lind, and G.S. Lakle, "Electromagnetic

Noise Studies of Severe Convective Storms,'' *Journal of Atmospheric Science,* 1971, vol. 28, pp. 436-48.

[15] Taylor, W.L., "Evaluation of an Electromagnetic Tornado-Detection Technique," *Preprints, Eighth Conference on Severe Local Storms,* American Meteorological Society, Boston, Mass., 1973, pp. 165-68.

[16] Eagleman, J.R., "Tornado Damage Patterns in Topeka, Kansas, June 8, 1966," *Monthly Weather Review,* 1967, vol. 95, no. 6, pp. 370-74.

[17] Brooks, E.M., "Tornadoes and Related Phenomena," *Compendium of Meteorology,* American Meteorological Society, Boston, Mass., 1951, pp. 673-80.

[18] Environmental Science Services Administration, *Tornadoes,* Washington, D.C., 1966, 15 pp.

[19] Galway, J.G., "The Topeka Tornado of 8 June, 1966," *Weatherwise,* 1966, vol. 19, no. 4, pp. 144-49.

[20] Budney, L.J., "Unique Damage Patterns Caused by a Tornado in Dense Woodlands," *Weatherwise*, April 1965, vol. 18, no. 2, pp. 75-77, 86.

[21] Reynolds, G.W., "A Common Wind Damage Pattern in Relation to the Classical Tornado," *Bulletin of the American Meteorological Society,* 1957, vol. 38, no. 1, pp. 1-5.

[22] Fujita, T.T., "The Lubbock Tornadoes: A Study of Suction Spots," *Weatherwise,* 1970, vol. 23, no. 4, pp. 160-73.

[23] U.S. Department of Commerce, *Climatological Data, National Summary,* U.S. Government Printing Office, Washington, D.C., 1950-74.

[24] U.S. Department of Commerce, *Storm Data,* U.S. Government Printing Office, Washington, D.C., 1950-74.

[25] U.S. Department of Commerce, "Tornado Occurrences in the United States," *Technical Paper No. 20,* U.S. Government Printing Office, Washington, D.C., 1953, 43 pp.

[26] Ludlam, F.H., "Severe Local Storms, A Review," *Meteorological Monographs, Severe Local Storms,* American Meteorological Society, Boston, Mass., 1963, vol. 5, no. 27, pp. 1-32.

[27] Byers, H.R. and L.J. Battan, "Some Effects of Vertical Wind Shear on Thunderstorm Structure," *Bulletin of American Meteorological Society,* 1949, vol. 30, pp. 168-75.

[28] Hitschfeld, W., "The Motion and Erosion of Convective Storms in Severe Vertical Wind Shear," *Journal of Meteorology,* 1960, vol. 17, pp. 270-83.

[29] Newton, C.W. and H.R. Newton, "Dynamical Interactions between Large Convective Clouds and Environmental Clouds with Vertical Shear," *Journal of Meteorology,* 1959, vol. 16, pp. 483-96.

[30] Charba, J. and L. Sasaki, "Structure and Movement of the Severe Thunderstorms of April 3, 1964, as Revealed from Radar and Surface Mesonetwork Data Analysis," *Technical Memorandum,* ERLTM-NSSL41, 1968, 47 pp.

[31] Byers, H.R., "Non-Frontal Thunderstorms," *Miscellaneous Report, No. 3,* Department of Meteorology, University of Chicago, 1942, 22 pp.

[32] Browning, K.A., "Airflow and Precipitation Trajectories within Severe Local Storms Which Travel to the Right of Winds," *Journal of Atmospheric Sciences,* 1964, vol. 21, pp. 634-39.

[33] Fankhauser, J.C., "Thunderstorms—Environment Interaction Determined by Aircraft and Radar Observation," *Monthly Weather Review,* 1971, vol. 99, no. 3, pp. 171-92.

[34] Fujita, T.T. and Hector Grandoso, "Split of a Thunderstorm into Anticyclonic and Cyclonic Storms and Their Motion as Determined from Numerical Model Experiments," *Journal of Atmospheric Sciences,* 1968, vol. 23, no. 3, pp. 416-39.

[35] Goldman, J.L., "The Role of the Kutta-Joukowski Forces in Cloud Systems with Circulation," *Technological Notes,* 1966, vol. 48-NSSL, no. 27, pp. 21-24.

[36] Rauscher, Manfred, *Introduction to Aeronautical Dynamics,* John Wiley and Sons, New York, N.Y., 1953, 664 pp.

[37] Sourbeer, R.H. and R.C. Gentry, "Rainstorms in Southern Florida," *Monthly Weather Review,* Jan. 1957, vol. 89, pp. 9-16.

[38] Stout, G.E., "Mesometeorological Systems from Dense Network Systems," presented in International Union of Geodesy and Geophysics (IUGG) meeting in Toronto, 1957.

[39] Byers, H.R. and R.R. Braham, *The Thunderstorm,* U.S. Government Printing Office, Washington, D.C., 1949, 287 pp.

[40] Fujita, T.T., "Analytical Mesometeorology: A Review," *Meteorological Monograph, Severe Local Storms,* American Meteorological Society, vol. 5, no. 27, 1963, pp. 77-122.

[41] Eagleman, J.R., V.U. Muirhead and Nicholas Willems, "Thunderstorms, Tornadoes and Damage to Buildings," *Research Report, HEW contract No. EC00303,* University of Kansas, 1970, 253 pp.

[42] Browning, K.A., "Some Inferences about the Updraft within Severe Local Storms," *Journal of Atmospheric Sciences,* vol. 22, no. 6, 1965, pp. 667-69.

[43] Dessens, H.J.J., "Severe Hailstorms Are Associated with Very Strong Winds between 6,000 and 12,000 M.," *The Physics of Precipitation, Geophysical Monograph,* no. 15, Washington, D.C., 1960, pp. 333-38.

[44] Browning, K.A. and T.T. Fujita, "A Family of Severe Local Storms—A Comprehensive Study of the Storms in Oklahoma on May 26, 1963," *Part I, Air Force Cambridge Research Laboratory,* Bedford, Mass., 1965.

[45] Bradbury, D.L. and T.T. Fujita, "Features and Motions of Radar Echoes on Palm Sunday, 1965," *SMRP Research Paper* no. 51, University of Chicago, 1966, 23 pp.

[46] Wills, T.G., "Characteristics of the Tornado Environment as Deduced from Proximity Soundings," *Preprints, Sixth Conference on Severe Local Storms,* American Meteorological Society, Boston, Mass., 1969, pp. 222-29.

[47] Lemon, L.R., "Formation and Emergence of an Anticyclonic Eddy within a Severe Thunderstorm as Revealed by Radar and Surface Data," *Preprints, Fourteenth Radar Meteorology Conference,* American Meteorological Society, Boston, Mass., 1970, pp. 323-28.

[48] Donaldson, R.J., Jr., G. Armstrong, A.C. Chemal, and M.J. Draus, "Doppler Radar Investigation of Air Flow and Shear within Severe Thunderstorms," *Preprints, Sixth Conference on Severe Local Storms,* American Meteorological Society, Boston, Mass., 1969, pp. 146-54.

[49] Segman, Ralph, "ESSA Doppler Radar System," *Weatherwise,* 1970, vol. 23, pp. 70-73, 83, 103.

[50] Browning, K.A. and F.H. Ludlam, "Air Flow in Convective Storms," *Quarterly Journal of Royal Meteorological Society,* 1962, vol. 8, pp. 117-37.

[51] Marwitz, J.D., "The Structure and Motion of Severe Hailstorms, Part I: Supercell Storms," *Journal of Applied Meteorology,* 1972, vol. 11, no. 1, pp. 166-79.

[52] Donaldson, R.J., Jr., "Radar Observations of a Tornado Thunderstorm in Vertical Section," *National Severe Storm Project Report, No. 8,* 1962, 21 pp.

[53] Bates, F.C., "Tornadoes in the Central United States," *Transaction, Kansas Academy of Science,* 1962, vol. 65, pp. 215-46.

[54] Wichmann, Helmut, "Uber das Vorkommen und Verhalten des Hagels in Gewitterwolden," *Meteorological Annual,* 1951, vol. 4, pp. 218-25.

[55] Newton, C.W., "Dynamics of Severe Convective Storms," *Meteo-*

rological Monographs, Severe Local Storms, American Meteorological Society, 1963, vol. 5, no. 27, pp. 33-58.

[56] Harrold, T.W., "A Note on the Development and Movement of Storms over Oklahoma on May 7, 1965," *National Severe Storms Laboratory Technical Memorandum, No. 29,* 1966, pp. 1-8.

[57] Achtemeier, G.L., "Some Observations of Splitting Thunderstorms over Iowa on August 25-26, 1965," *Department of Meteorology Report No. 69-4,* Florida State University, 1969, 17 pp.

[58] Pautz, M.E., "Severe Local Storm Occurrences, 1955-1967," *ESSA Technical Memorandum WBTM RCST 13,* Office of Meteorological Operations, Silver Springs, Md., 1969, 77 pp.

[59] Brooks, E.M., "The Tornado Cyclone," *Weatherwise,* 1949, vol. 2, pp. 23-33.

[60] Vonnegut, Bernard, "Electrical Theory of Tornadoes," *Journal of Geophysical Research,* 1960, vol. 65, pp. 203-12.

[61] Huff, F.A., H.W. Hiser, and S.G. Bibler, "Study of an Illinois Tornado Using Radar Synoptic Weather and Field Survey Data," *Report of Investigation Number 22,* Illinois State Water Survey, Urbana, Ill., 1954, 73 pp.

[62] Darkow, G.L., "Periodic Tornado Production by Long-Lived Parent Thunderstorms," *Preprints, Seventh Conference on Severe Local Storms,* American Meteorological Society, Boston, Mass., 1971, pp. 214-27.

[63] Eagleman, J.R., V.U. Muirhead, and Nicholas Willems, *Thunderstorms, Tornadoes and Damage to Buildings,* Environmental Publications, Lawrence, Kan., 1972, 279 pp.

[64] League, L.D., "Wind Flow Simulations Around and within Double Vortex Thunderstorm Cells," M.A. Thesis, University of Kansas, Lawrence, Kan., 1971, 90 pp.

[65] Nunley, R.E., "Living Maps of the Field Plotter," *Commission of College Geography Technical Paper No. 4,* Association of American Geographers, Washington, D.C., 1971, 155 pp.

[66] Newton, C.W. and J.C. Fankhauser, "On the Movements of Convective Storms, Emphasis on Size Discrimination in Relation to Water-Budget Requirements," *Journal of Applied Meteorology,* 1964, vol. 3, pp. 651-68.

[67] Harrison, H.T. and W.I. Orendorff, "Prefrontal Squall Lines," United Air Lines, Meteorology Dept., *Circulation Memo #29,* 1941, pp. 1-8.

[68] Reynolds, G.W. "Venting and Other Building Practices as Practical Means of Reducing Damage from Tornado Low Pressures," *Bulletin*

of American Meteorological Society, Jan. 1958, vol. 39, no. 1, pp. 14-20.

[69] Eagleman, J.R. and V.U. Muirhead, "Observed Damage from Tornadoes and Safest Location in Houses," *Preprints, Seventh Conference on Severe Local Storms,* American Meteorological Society, Boston, Mass., Oct. 1971, pp. 171-77.

[70] Eagleman, J.R., V.U. Muirhead, Nicholas Willems, "Thunderstorms, Tornadoes and Damage to Buildings," *Research Report, Contract No. EC00303,* University of Kansas, Lawrence, Kan., Dec. 1971, 290 pp.

[71] Muirhead, V.U. and J.R. Eagleman, "Laboratory Compressible Flow Tornado Model," *Preprints Seventh Conference on Severe Local Storms,* American Meteorological Society, Boston, Mass., Oct. 1971, pp. 284-91.

[72] Muirhead, V.U., "Compressible Vortex Flow," *American Institute of Aeronautics and Astronautics Paper No. 73-106,* Eleventh Aerospace Sciences meeting, New York, N.Y., Jan. 1973.

[73] Muirhead, V.U. and J.R. Eagleman, "Tornado Vortex Air Flow," *Preprints, Eighth Conference on Severe Local Storms,* American Meteorological Society, Boston, Mass., 1973, pp. 213-28.

[74] Melarango, M.G., *Tornado Forces and Their Effects on Buildings,* Kansas State Printing Service, Manhattan, Kan., 1968, 51 pp.

[75] Lamb, Horace, *Hydrodynamics,* 6th publication, Dover Publication, New York, N.Y., 1932.

[76] Lewellen, W.S., W.S. Burns, and H.J. Strickland, "Transonic Swirling Flow," *American Institute of Aeronautics and Astronautics, Paper No. 7,* 1290-12970, New York, N.Y., 1969.

[77] Chang, C.C., "Recent Laboratory Model Study of Tornadoes," *Preprints, Sixth Conference on Severe Local Storms,* American Meteorological Society, Boston, Mass., Apr. 1969, pp. 244-52.

[78] Eagleman, J.R., "A Double Vortex Thunderstorm Model," *Preprints, Seventh Conference on Severe Local Storms,* American Meteorological Society, Boston, Mass., 1971, pp. 177-78.

[79] Pope, Alan, *Basic Wind and Airfoil Theory,* 1st ed., McGraw-Hill Publications, New York, N.Y., 1951.

[80] Liepmann, H.W. and Anatol Roshko, *Elements of Gasdynamics,* 3rd pub., John Wiley & Sons, Inc., New York, N.Y., 1960.

[81] *Boeing Report DG-32322 TN,* "737 Engine Inlet Vortex Dissipator Development Program," Boeing Company, Seattle, Wash., Dec. 1969.

[82] Turner, T.R., "Windtunnel Investigation of a 3/8 Scale Automobile over a Moving Belt Ground Plane," *NASA TN D-4229*, Nov. 1967.

[83] Matthews, J.T. and W.F. Barnett, "Wind Tunnel Tests of a Scale Model Railroad Automobile Rack Car," *U.S. Department of Transportation, PB-180198*, June 1968.

[84] Galway, J.L., "The Lifted Index as a Predictor of Latent Instability," *Bulletin of American Meteorological Society*, 1956, vol. 36, pp. 528-29.

[85] Fawbush, E.J. and R.C. Miller, "A Mean Sounding Representative of the Tornado Air Mass," *Bulletin of American Meteorological Society*, 1952, vol. 33, no. 7, pp. 303-7.

[86] Fawbush, E.J. and R.C. Miller, "The Types of Air Masses in which North American Tornadoes Form," *Bulletin of American Meteorological Society*, 1954, vol. 35, pp. 154-65.

[87] Showalter, A.K., "A Stability Index for Forecasting Thunderstorms," *Bulletin of American Meteorological Society*, 1953, vol. 34, p. 250.

[88] Darkow, G.L., "The Total Energy Environment of Severe Storms," *Journal of Applied Meteorology*, 1968, vol. 7, no. 2, pp. 199-205.

[89] Fujita, T.T., "Results of Detailed Synoptic Studies of Squall Lines," *Tellus*, 1955, vol. 7, pp. 405-436.

[90] Crumrine, H.A., "The Use of the Horizontal Temperature Advection, the 850-500 mb Thickness and 850-500 mb Shear Wind as an Aid in Severe Local Thunderstorm Forecasting," USWB Library, Kansas City, Mo., 1965.

[91] Gray, W.M., "Hypothesized Importance of Vertical Wind Shear in Tornado Genesis," *Preprints, Sixth Conference on Severe Local Storms*, American Meteorological Society, Boston, Mass., 1969.

[92] Wills, T.G., "Characteristics of the Tornado Environment as Deduced from Proximity Soundings," *Preprints, Sixth Conference on Severe Local Storms*, American Meteorological Society, Boston, Mass., 1969.

[93] Newton, C.W. and J.C. Fankhauser, "On the Movements of Convective Storms, with Emphasis on Size Discrimination in Relation to Water-budget Requirements," *Journal of Applied Meteorology*, 1964, vol. 3, no. 6, pp. 651-68.

[94] Beebee, R.G., "Tornado Proximity Soundings," *Bulletin of American Meteorological Society*, 1958, vol. 38, pp. 195-201.

[95] Dryden, J.L. and G.C. Hill, "Wind Pressures on Structures," *U.S. National Bureau of Standards*, 1926, Science Paper 523, vol. 3.

[96] Ning Chien, Yin Feng, Hung-Ju Wang, and Tien-To Siao, "Wind Tunnel Studies of Pressure Distribution on Elementary Building Forms," *Iowa Institute of Hydraulic Research*, State University of Iowa, 1951.

[97] Salter, C.A., "Wind Loadings on Flat Roof Buildings," *Engineering*, 1958, vol. 186, no. 4832, pp. 508-10.

[98] Irminger, J.O.V. and C. Nøkkentved, *Wind Pressure on Buildings Experimental Researches*, Ingen Ividenskabelige Skrifter (2nd Series), Copenhagen, 1936.

[99] Jensen, Martin, "The Model Law for Phenomena in Natural Wind," *Ingeniøren*, 1958, vol. 2, no. 4, pp. 121-28.

[100] Haddon, V.D., "The Use of Wind Tunnel Models for Determining the Wind Pressure on Buildings," *Civil Engineering and Public Works Review*, 1960, vol. 55, no. 645, p. 500.

[101] ASCE, "Wind Forces on Structures," *Final Report of Task Committee on Wind Forces*, American Society of Civil Engineers Transaction, 126, part II, 1961, pp. 1124-98.

[102] MRI, "Wind Effect on a Flat Roof," *Final Report Midwest Research Institute Project No. 2815-P*, 1965.

[103] Gibbs, P.D., "The Effect of Oscillatory Winds on Flat Roofs," Thesis, University of Kansas, Lawrence, Kan., 1967.

[104] *Wind Effects on Buildings*, Proceedings, International Research Seminar, University of Toronto Press, 1968, vol. 1, 772 pp.

[105] *Wind Effects on Buildings*, Proceedings, International Research Seminar, University of Toronto Press, 1968, vol. 2, 461 pp.

[106] Pope, Alan and J.J. Harper, *Low-Speed Wind Tunnel Testing*, John Wiley and Sons, 1966, 325 pp.

[107] Hoyle, R.J., *Wood Technology in the Design of Structures*, Mountain Press Publishing Co., Missoula, Mont., 1973.

[108] Lemon, L.R., "Thunderstorm Wake Vortex Structure and Aerodynamic Origin," *NOAA technical memo ERL NSSL-71*, 1974, pp. 17-43.

[109] Donaldson, R.L., Jr., "Doppler Radar Evidence for Anticyclonic Circulation in a Severe Convective Storm," *Preprint Volume, Eighth Conference on Severe Local Storms*, American Meteorological Society, Boston, Mass., 1973, pp. 48-50.

[110] Schlichting, Hermann, *Boundary Layer Theory*, trans. J. Kestin, McGraw-Hill, New York, N.Y., 1955.

[111] White, F.M., *Viscous Fluid Flow*, McGraw-Hill, 1974, pp. 423-25.

[112] Fujita, T.T., "Proposed Mechanism of Suction Spots Accompanied

by Tornadoes,'' *Preprints, Seventh Conference on Severe Local Storms,* American Meteorological Society, Boston, Mass., 1971, pp. 208-13.

[113] ''Uniform Building Code,'' International Conference of Building Officials, 1973.

[114] ''Minimum Property Standards,'' Federal Housing Administration, HUD, Jan. 1965.

[115] *Research on Short-Term Weather Phenomena,* Hearings before the Subcommittee on Space Science and Applications, Committee on Science and Astronautics, U.S. House of Representatives, 93rd Congress, Nov. 6, 7, and 9, 1973, U.S. Government Printing Office, 1974, 329 pp.

[116] Melarango, M.C., ''Outdoor Tornado Shelters for Residential Areas,'' *K-State Printing Service,* Kansas State University, Manhattan, Kan., 1968, 36 pp.

Index

Index

About the Authors

Joe R. Eagleman is Professor of Meteorology at the University of Kansas. He received the Ph.D. in meteorology from the University of Missouri in 1963. Professor Eagleman is the director of the Atmospheric Science Laboratory at the Space Technology Center, University of Kansas, where he has directed a variety of research programs. He is the author of a number of technical articles which have been published in magazines such as *Monthly Weather Review, Journal of Applied Meteorology, Advances of the American Astronomical Society,* proceedings of the *International Symposium on Remote Sensing of Environment, International Journal of Agricultural Meteorology,* and *Atmospheric Environment.*

Vincent U. Muirhead is Associate Professor of Aerospace Engineering at the University of Kansas. He is a graduate of the U.S. Naval Academy and received the degree of Aeronautical Engineer from the California Institute of Technology. Professor Muirhead has been a consultant with Black and Veatch, Consulting Engineers since 1964.

Nicholas Willems is Chairman of the Civil Engineering Department at the University of Kansas. He received the Ph.D. degree in engineering from the University of Kansas in 1963. Before coming to the United States, Dr. Willems worked in various engineering capacities in South Africa and Holland. He has published articles in many engineering journals such as *Journal of Applied Mechanics, Journal of Engineering for Industry,* and *ASCE Journal, Structural Division.*